Making Better Environmental Decisions

Making Better Environmental Decisions
An Alternative to Risk Assessment

Mary O'Brien

published in association with the Environmental Research Foundation

The MIT Press
Cambridge, Massachusetts
London, England

Set in Sabon by The MIT Press.
Printed and bound in the United States of America.
Printed on chlorine-free, recycled paper.

Library of Congress Cataloging-in-Publication Data

O'Brien, Mary, 1945–
Making better environmental decisions: an alternative to risk assessment / Mary O'Brien
p.   cm.
Includes bibliographical references and index.
ISBN 0-262-15051-4 (cloth: alk. paper)—ISBN 0-262-65053-3 (pbk.: alk. paper)
1. Environmental risk assessment. 2. Environmental policy—decision making. I. Title.
GE145.O27 2000
333.7'14—dc21   99-056868

Contents

Foreword

This book represents a watershed event in environmental thinking—a paradigm shift away from risk assessment and toward the principle of precautionary action. For decision makers, improved new tools are at hand.

What is risk assessment, and why does it need to be supplanted? During the late 1960s it slowly became clear that many modern technologies had far surpassed human understanding, giving rise to by-products that were dangerous, long-lived, and almost completely unanticipated. A book-length report issued by the White House in November of 1965 (John W. Tukey et al., Restoring the Quality of Our Environment) began with a letter from President Lyndon Johnson, who wrote: "Ours is a nation of affluence. But the technology that has permitted our affluence spews out vast quantities of wastes and spent products that pollute our air, poison our waters, and even impair our ability to feed ourselves." The 1965 White House report identified numerous major sources of environmental contamination: municipal and industrial sewage, animal wastes, municipal solid wastes, mining wastes, and "unintentional releases" (which included automobile exhaust, smokestack emissions, pesticidal mists, and agricultural chemicals draining into waterways, among other things). The main report contained "subpanel reports" on soil contamination, the potential for global warming by carbon dioxide, the effects of chlorinating wastes, the health effects of environmental pollution, and "the effects of pollutants on organisms other than man."

As the 1960s progressed, new reports from governments and from independent researchers brought a string of unpleasant surprises indicating environmental decline.

By the mid 1970s it was clear that industrial technology had an immense, unpredictable dark side. Technical mastery of natural forces was leading not to safety and well being but to a random shredding of the biological platform upon which we all stand.

In response to a decade of disturbing revelations, a vast "environmental movement" developed, made up of citizens concerned about one place or another—their dinner table, the playground in their neighborhood, the river running through their city (often the source of their drinking water). They demanded reforms. Congress reacted by writing laws the size of the Los Angeles telephone book, creating new agencies and departments to issue enforceable regulations.

The nation's new risk-based regulatory system was founded on three unspoken assumptions, summarized by Theodore Taylor and Charles Humpstone in *The Restoration of the Earth* (Harper & Row, 1973). Paraphrased, these assumptions are as follows:

• .Humans can "manage" the environment by deciding how much of any destructive activity the Earth (or any portion of the Earth) can safely absorb without harm. Scientists call this the "assimilative capacity" of an ecosystem or a human being or a population of fish. According to this assumption, scientists can reliably decide how much damage a river or a human or the Florida panther can absorb without suffering irreversible harm. The purpose of every risk assessment is to predict the limits of this "assimilative capacity."

• Once an ecosystem's "assimilative capacity" for a particular toxicant (or destructive activity, such as road building in grizzly habitat) has been decided, then we can and will impose limits so that irreversible harm will not occur. We will set restrictions, river by river, forest by forest, factory by factory, chemical by chemical, everywhere on the planet, so that the total, cumulative effects do not exceed the "assimilative capacity" of the Earth or any of its ecosystems or inhabitants.

• We already know which substances and activities are harmful and which are not; or, in the case of substances or activities that we never suspected of being harmful, we will be warned of their possible dangers by traumatic but sub-lethal shocks that alert us to the danger before it is too late.

Unfortunately, with the benefit of 30 years' hindsight we now know that all three assumptions are dead wrong.

For 30 years we have relied on a risk-based environmental-protection apparatus based on these incorrect assumptions. As a result, the world's ecosystems and its human inhabitants have suffered major damage. Think of ozone depletion, global warming, acid rain, the lead poisoning of tens of millions of children, mercury buildup in the atmosphere and subsequently in fish, industrial poisons (such as PCBs) measurable throughout the vastness of the oceans, the rising incidence of many cancers (including brain cancers, lymphomas, and childhood cancers), escalating immune system disorders (including asthma and diabetes), the rising incidence of nervoussystem diseases (such as Parkinson's Disease and Lou Gehrig's Disease), coral reefs dying worldwide, and numerous species gone extinct. This list could be extended readily as new evidence of harm pours in.

All this damage has resulted from our "innocent until proven guilty" approach to destructive activities, including the massive ongoing release of industrial poisons. If our risk assessments told us some activity would cause only "acceptable damage," we plunged ahead on that basis. Now we wake up to find that the cumulative effects of our risk-based decisions have severely degraded many of the planet's biological systems.

On the positive side, a new approach is now available, giving decision makers improved tools, and giving everyone reason for hope. The new approach is based on a "Principle of Precautionary Action." In a nutshell, as explained by Carolyn Raffensperger and Joel Tickner in *Protecting Public Health and the Environment* (Island, 1999), that principle says the following:

When an activity raises threats of harm to human health or the environment, precautionary measures should be taken even if some cause and effect relationships are not fully established scientifically. In this context the proponent of an activity, rather than the public, should bear the burden of proof.

The process of applying the Precautionary Principle must be open, informed and democratic and must include potentially affected parties. It must also involve an examination of the full range of alternatives, including no action.

The basic idea? Faced with evidence of possible harm, and faced with scientific uncertainty, we should make decisions based on familiar maxims: An ounce of prevention is worth a pound of cure. Look before you leap. Do unto others as you would have others do unto you. Better safe than sorry.

If you have reason to believe that your building may be on fire, do you estimate the probability that the damage will be "acceptable" and wait until you see flames shooting into the sky? Or do you take precautionary action based on incomplete evidence and call the fire department?

The principle of precautionary action is not a perfect guide. We have much to learn about how to apply it in various situations. Mary O'Brien's book makes an important contribution toward filling in the most important blanks—the part about examining the full range of alternatives. Risk assessment might be one of the many ways in which we choose to examine a range of alternative actions; however, when used in this comparative way, risk assessment bears almost no resemblance to the old-style use of risk assessment to divine the limits of the Earth's "assimilative capacity."

This book signals a new day for decision makers. It provides ethical and intellectual support and guidance to those who are making their way down this uncharted path of precautionary action. And it provides hope for us all.

Peter Montague
Environmental Research Foundation
Annapolis, Maryland

Preface

This book has arisen out of 16 years of facing risk assessments. I first met risk assessment, new Botany Ph.D. in hand, when citizens' groups in Oregon were attempting to move the Pacific Northwest Region of the U.S. Forest Service off its addiction to herbicide spraying in clearcuts and tree plantations. The Forest Service was using risk assessment to claim that these herbicides, including dioxin-contaminated 2,4,5,-T, would cause no harm. Oregon citizens' groups working for pesticide reform, and many tree planters, knew that the herbicides weren't safe and that non-chemical alternatives for preventing and controlling unwanted vegetation had been successfully used in Northwest forestry, but the Forest Service didn't want to talk about these alternatives. After all, their hired risk assessors had showed them that their herbicide use was safe.

As a scientist with a fierce trust in the physical world, I first contended that if risk assessments were constructed with more attention to reality and scientific evidence they would bring wiser decision making. I began to work to make the Forest Service's risk assessments more reality-based.

In subsequent years, I watched dozens of risk assessments being constructed and manipulated to justify or defend hazardous activities and substances: dacthal contamination of groundwater in the onion-growing area of eastern Oregon; aerial spraying of carbaryl over Oregon cities to eradicate gypsy moth; use of chlorine to bleach pulp in mills on the polluted and radioactive Columbia River; construction of dams on salmon-bearing streams in southern Oregon; the destruction of old-growth forests and their dependent, endangered animal species; the construction of more roads across critical wildlife migratory routes on the rim of Hells Canyon; the shrinking of the grizzly's habitat in Montana.

I came across risk assessments in all the work settings I inhabited: as a Staff Scientist with the Northwest Coalition for Alternatives to Pesticides (Eugene, Oregon), as a Staff Scientist with the U.S. office of Environmental Law Alliance Worldwide (also in Eugene), as an Assistant Professor in the graduate Environmental Studies Program of the University of Montana (Missoula), as a member of the Board of Directors of the North American Regional Office of Pesticide Action Network, International (San Francisco), as an Ecosystem Policy Analyst with the Hells Canyon Preservation Council (LaGrande, Oregon), as a Staff Scientist for Environmental Research Foundation (Annapolis, Maryland), and as a member of international, state, and local technology, school grounds, river, and groundwater committees.

I even came across one risk assessment that candidly faced the implications of cumulative impacts, multiple species, and lack of information. But by then I had been disabused of my earlier faith in the role played by either science or candor in the risk-assessment process. Certain patterns in this process had gradually become clear to me (though they had earlier been clear to some other people): First, the process was inherently a fiction. Second, it was asking the wrong question. It was asking how much damage was safe, rather than how little damage was possible. Third, it was promoted most adamantly by those who stood to gain the most from it—i.e., those who were proposing to undertake activities that would be toxic or have other ill effects on the environment. Finally, risk-assessment, cost-benefit, and risk-benefit processes almost never acknowledged or seriously considered the numerous less-hazardous alternatives employed or understood by thoughtful and experienced farmers, foresters, businesses, government employees, or citizens.

Probably the most useful question I know is "What are the alternatives?" It occurred to me that the alternative to risk assessment is the same alternative most of us use in our daily lives: Spread out on the table all reasonable options for a decision, and look at the pros and cons of each of the options. I met others—for example, Carol Van Strum in a run-down farmhouse in coastal Oregon, and Nicholas Ashford at the Massachusetts Institute of Technology—who had the same thoughts. Peter Montague of the Environmental Research Foundation asked me to convey these thoughts in a book.

This book is about the only alternative I know to risk assessment, namely *alternatives assessment*. It is a simple and sensible alternative, but it is resisted mightily because consideration of options is a threat to business as usual and business as planned—and, often, to established power arrangements.

This book is grounded in the social experience of risk assessment, rather than in the theory, history, critical assessment, or possible permutations of risk assessment. It is written for laypersons as well as for practitioners, purveyors, and consumers of risk assessment.

I believe it is largely laypersons who will challenge and reduce the power of risk assessment, as so many laypersons already have. These people resist risk assessment because they suffer, lose, and even die from activities pronounced "safe" by a risk-assessment process or "acceptable" on the basis of risk assessment. They suffer, lose, and die from their society's failure to entertain the possibilities of more socially and environmentally sound ways of producing, consuming, and regulating.

I am also articulating alternatives assessment for risk assessors, because many of them can help create or insist upon opportunities to present their assessments of risks within larger alternatives assessments.

Some warnings about this book are in order. The warnings don't compromise the book; they merely require readers to adapt it to their particular circumstances.

Warning 1: If you practice risk assessment and you believe that you are thereby helping to protect the environment and the public health, you may feel criticized or dismissed by the ideas in this book. I believe that many of you are working mightily within the risk-assessment framework to bring more evidence, candor, and realism to the risk-assessment process and to environmental-protection efforts. Many of you are working courageously to include the public in your risk-assessment processes. But I urge you to contemplate whether the risk-assessment frame is large enough for you. Are you being allowed the room to learn about and calculate the pros and cons of all reasonable alternatives to the activity whose risks you are calculating? Risk assessment is, to some degree, dependent on the motives, assumptions, data bases, and intent of the risk assessors, but not as much as many people would hope. This is because risk assessment is fundamentally played out on one highly structured turf. The game played on this turf is to estimate, with more or less information and candor, how much of a hazardous activity is safe, insignificantly harmful, or acceptable. Alternatives assessment is a different game, played on different turf. The goal is to gather information from many people about the pros and cons of a variety of different paths to a goal. The goal itself is conceived as broadly as possible. If you are a skilled, honest, committed risk assessor, you will be able to do much more to evoke needed information for good decisions about public health and the environment over on the turf of alternatives assessment.

Warning 2: The examples in this book draw heavily, although not exclusively, from the western United States, because that's my home. Some things are unique in the West, but risk assessment is not one of them. As far as risk assessment goes, Oregon is Louisiana, Siberia, the Netherlands, Australia, and Chile. I know: I've been to those places to see risk assessments, and the risk assessments look just like those in the Western United States. Likewise, the examples in the book draw heavily from the world of toxic chemicals. But risk assessment is the same process, whether it's undertaken for toxic chemicals, radiation, fish harvests, noisy workplaces, road construction, or mining gold. In every risk assessment, risk assessors talk about some of the impacts of some activity or activities on humans and/or the environment, and then they or whoever hired them talks about what amount of that activity is safe or acceptable.

Warning 3: This book is obviously grounded in advocacy for the environment, public health, and democracy. However, that doesn't compromise its objectivity, scientific validity, or depth. While I passionately defend environmental and public health, I also passionately defend realism, science, excellence, and truth, to the best of my understanding. While I am an advocate, I also remain a scientist.

I wish to be made aware of any errors, and I look forward to a continuing dialogue.

Acknowledgments

I thank, first of all, the citizens, citizen organizations, and tribes of this and other countries who have worked valiantly to save communities and ecosystems—other species, their families, and themselves—from the destruction that attends unnecessary activities that supposedly pose "insignificant" risks. They showed me *why* risk assessment has to be replaced with the honesty, reality, and dignity of alternative ways of behaving. Their proposals for better ways of behaving show how risk assessment will be replaced.

I thank the Environmental Research Foundation for employing me while I took too long to write this book. The information in dozens of issues of Peter Montague's *Rachel's Environment & Health Weekly* was for me, as it is for much of the environmental movement in this country, indispensable.

The Bullitt Foundation, the CS Fund, and the C. S. Mott Foundation provided financial support to the Environmental Research Foundation for this endeavor.

I particularly want to thank the groups who co-sponsored workshops on alternatives assessment: The Environmental Health Coalition (San Diego), the JSI Center for Environmental Health Studies, the Boston University School of Public Health and its Department of Environmental Health, and the Washington Toxics Coalition.

I also thank Nancy Adess of Point Reyes, California, for her skillful editing, and Andrea Fearneyhough, Becky Paul, and especially Maria B. Pellerano (all of the Environmental Research Foundation), who spent many months locating essential pieces of information, putting the manuscript into final form, checking references, and verifying facts.

The ideas in this book first formed during my years as a staff scientist with Northwest Coalition for Alternatives to Pesticides, and I am grateful for the support they offered for that exploration and their ongoing response to my requests for information.

It was Carol Van Strum who first warned me that "making risk assessments better" was not going to work. She merely smiled when I came to the same conclusion; she graciously didn't say "I told you so."

For years, the following people have played constant Devil's Advocate to my ideas about risk assessment, forcing me to think harder and consider more aspects: Peter Montague, Monica Moore, Joel Pagel, Vic Sher, Joel Tickner, Carol Van Strum, and Tom Webster. I thank Peter, both Joels, Vic, and Carol for combing through a draft manuscript. In addition, Russell Boulding provided helpful comments.

John and Ellen Hough let me stay for a week, alone, in their house among John's constructed wetlands to review the first draft. There I had space and silence in which to review more of my own words than I cared to.

I am not able to list by name all the people who responded to my requests for documents, examples, and obscure numbers, dates, and names, but these requests never were ignored. Their willingness to help was a source of inspiration.

I thank my family (John, Josh, and O'B) for not letting me think about risk assessment *all* the time, and for their acceptance of me when I thought about it too much. O'B always made me laugh, which ranks among the most fundamental of all sources of support.

Finally, I'm grateful to the Earth, because it is the ultimate source of information about whether risk assessment is wise. Earth seems to be calling out for alternatives to all the "insignificant" risks to which we have been subjecting it.

Mary O'Brien
Eugene, Oregon

Introduction

This book is based on the understanding that it is not acceptable for people to tell you that the harms to which they will subject you and the world are safe or insignificant. You deserve to know good alternatives to those harms, and you deserve to help decide which alternative will be chosen.

Underlying this book, however, is a less explicitly stated personal belief, namely that we humans will never dredge up enough will to alter our habitual, destructive ways of behaving toward each other and the world unless we simultaneously employ information *and* emotion *and* a sense of relationship to others—other species, other cultures, and other generations. Using information while divorced from emotion and using information while insulated from connection to a wide net of others are how destruction of the Earth is being accomplished.

Risk assessment of narrow options is a classic example of using certain bits of information in such a way as to exclude feeling and to artificially sever connections of parts to the whole. Risk assessment rips you (and others) out of connection to the rest of the world and reduces you (if you are even considered at all in the risk assessment) to a number. You are then consigned to damage or death or "risk," depending on how your number is shuffled around in models, assumptions, and formulas and during "risk management."

Assessment of the pros and cons of a range of reasonable *alternatives* allows the connections to remain. The cultural emotions connected to a given alternative, for instance, can be a pro or a con, and may be both, depending on which sector of the community you inhabit. An advantage or a disadvantage of a given alternative can be social, religious, economic, scientific, or political.

Risk assessment is one of the major methods by which parts (corporations such as Monsanto or Hyundai, "private landowners," industrial nations) can act on their wants at the expense of wholes (e.g., whole communities and countries, or the seventh generation from now) without appearing to be doing so. Risk assessment lets them appear simply "scientific" or "rational" as they numerically estimate whether or how many deaths or what birth defects will be caused, and ignore other regions of human experience that also matter to people.

Always, some groups of humans will be trying to exercise their power at the expense of the whole. Decisions arrived at by risk assessment can be homicidal, biocidal, and suicidal, but they are made every day. Risk assessment is a premier process by which illegitimate exercise of power is justified. The stakes of installing alternatives to risk assessment, therefore, are the whole Earth (just as are the stakes of fashioning democratic control over corporations, or of requiring changes in behavior of those who have wreaked irreparable damage).

Installing alternatives assessment is one step in the struggle to use information, feeling, and a sense of relationship to others to stop socio-environmental madness.

This book shows how citizens (including scientists, activists, parents, government employees, teachers, physicians, judges, factory managers, farmers, and youth) can stop the processes by which unnecessary, inappropriate behavior is passed off as "acceptable" through risk assessment. It shows how to replace risk assessment with alternatives assessment.

While the entire book makes a comprehensive case for alternatives assessment, using real-world examples, most chapters can be read separately, as each concerns a specific aspect of the issue.

It is the goal of this book to provide citizens, risk assessors, and decision makers with both the reasons and the confidence to replace risk assessment with alternatives assessment.

I

What Is Wrong with Risk Assessment?

1

Goal: Replace Risk Assessment with Alternatives Assessment

The goal of this book is to help replace risk assessment of a narrow range of options with public assessment of a broad range of options. The book is based on a set of values, principles, and understandings, as well as on scientific and factual information.

Imagine a woman standing by an icy mountain river, intending to cross to the other side. A team of four risk assessors stands behind her, reviewing her situation. The toxicologist says that she ought to wade across the river because it is not toxic, only cold. The cardiologist says she ought to wade across the river because she looks to be young and not already chilled. Her risks of cardiac arrest, therefore, are low. The hydrologist says she ought to wade across the river because he has seen other rivers like this and estimates that this one is not more than 4 feet deep and probably has no whirlpools at this location. Finally, the EPA policy specialist says that the woman ought to wade across the river because, compared to global warming, ozone depletion, and loss of species diversity, the risks of her crossing are trivial.

The woman refuses to wade across. "Why?" the risk assessors ask. They show her their calculations, condescendingly explaining to her that her risk of dying while wading across the river is one in 40 million.

Still, the woman refuses to wade across. "Why?" the risk assessors ask again, frustrated by this woman who clearly doesn't understand the nature of risks.

The woman points upstream, and says "Because there is a bridge."

The risk assessors in this story are evaluating the risks of only one option: wading across an icy river. The woman is evaluating her alternatives, one of which involves crossing the river on a bridge. The woman doesn't really care whether getting wet in the icy stream will kill her or not, because it doesn't make sense to her to even become chilled in light of her options. This is the fundamental difference between risk assessment and alternatives assessment.

Now contemplate another story—one in which risk assessment takes place after, rather than before, a hazardous activity has begun.

Imagine a company whose air emissions coalesce during rain into heavy, baseball-size clumps. The balls hurt when they hit someone on the head. Occasionally the pollution balls cause concussions, but they rarely kill someone.

A team is hired by the company (or the county, or the federal government) to calculate how often someone's head is hit, how soon the pain generally goes away, how often a concussion is mild and how often severe, and which people seem particularly susceptible to death when hit by the pollution balls.

This team would be doing a risk assessment.

You might say to yourself that this is an absurd story. People would not simply stand by and do a risk assessment in such a situation. The company would not be allowed to emit pollutants that could ball up. They would be forced to change their production practices so pollution balls wouldn't be hitting people in the head.

When we do risk assessments, however, we do the precise equivalent of this, "standing by" and analyzing possible damages while people (company bosses, landowners, farmers, consumers, legislators) and the materials they use, create, and discard (in factories, on farms and ranches, in forests, in towns, in cities, at government installations) unnecessarily inflict stress, harm, or even death on both living things and natural processes. Thus, humans—ourselves, our children, and future generations—are affected, along with animal and plant life and the natural elements (such as groundwater and the ozone layer) that help support life. The activities that inflict harm include emitting toxic chemicals or particulates,[1] draining an underground water supply, allowing sediments to wash into a river, erecting dams that block salmon from returning home to spawn, draining wetlands to build developments, or killing trees with acid rain.

The process of estimating damages that may be occurring, or that may occur if an activity is undertaken, is called risk assessment.

Example: A large incinerator in East Liverpool, Ohio, burns hazardous wastes. Because the incinerator doesn't completely burn all of the toxic chemicals and doesn't burn toxic metals that are present in the wastes, it emits some highly toxic chemicals and metals into the air. The incinerator is located 400 yards from an elementary school. This school is up on a bluff, approximately level with the incinerator's stack. The toxic chemicals that leave the stack are carried in the wind over to the school, where the children are hit by them every day. They breathe the toxic chemicals into their lungs, they absorb them on their skin, they pick some up on their fingers in the schoolyard, and they ingest some whenever they stick their fingers in their mouths or eat their food.

Industry consultants and government employees calculate how much WTI, the corporation that runs the incinerator, shall be allowed to hit the elementary school children with toxic chemicals and metals (Hazardous Waste Facility Approval Board 1984; EPA Deputy Administrator Robert Sussman, letter to Terri Swearingen of Tri-State Environmental Council, July 16, 1993). They have decided, for instance, that the incinerator's managers can annually send 9400 pounds of lead, 2560 pounds of mercury, and 157,400 pounds of fine particles out of the 150-foot incinerator stack (Montague 1992a).

These people are doing risk assessment.

Most of the activities assessed in risk assessments produce some commercial benefit, or supposedly "solve" some problem, such as what to do with toxic wastes. The real or perceived benefits are generally considered justification for the activity being evaluated. For instance, providing jobs or competing in the global economy may be a benefit assumed for the commercial facilities that are bringing their toxic wastes to the WTI incinerator. An assumed benefit of the incinerator itself may be that it avoids the construction of new hazardous-waste landfills. The risk assessment will then focus on particular potential "risks" caused by incinerator operations, such as asthma or cancer from the toxic chemicals and metals escaping from the incinerator stack.

Thus, a risk-benefit or a cost-benefit analysis is implicit in a decision-making process involving risk assessment, even if the actual existence or extent of the benefits (e.g., jobs or waste disposal) is not explicitly examined in a formal risk-benefit or cost-benefit analysis. (See chapter 10.)

The thinking and the processes of risk assessment currently pervade policy decision making throughout many industrialized societies. The most basic, unstated goal of risk assessment, however, is to provide permission for undertaking some amount or form of the activity whose risks are being assessed.

It would be more helpful to the school children of East Liverpool if, instead of simply writing permits for "acceptable" exposure of the children to lead, mercury, and particulates from the WTI incinerator, we encouraged reduction of toxic-waste production by those industries that haul their wastes to the WTI incinerator. Similarly, we might be more helpful to the children if we worked with industries to help them develop and use alternative technologies that don't depend on using toxic chemicals. We could develop laws that require industries to keep their toxic wastes to themselves (for instance, by storing the wastes on their property in large, above-ground containers). Having to store their own wastes on their own property would give the companies incentives to reduce their production of wastes and to develop technologies that detoxify certain toxic chemicals.

If industry consultants, the government employees, or the communities and states involved in the WTI case examined the risks and benefits of these potentially helpful options alongside the risks and benefits of the proposed WTI incinerator operations, they would be doing alternatives assessment.

Fundamental Principles

Before plunging into my arguments, I want to lay out the principles on which both this book and the concept of alternatives assessment are based. These principles, or bases for conduct, come out of particular values and biases. (For each, I have indicated how risk assessment is based on different values and biases.)

Adherence to the following principles requires decision making based on diverse public participation and consideration of numerous social and technical options. In other words, the following principles require public-based alternatives assessment.

1. It is not acceptable to harm people when there are reasonable alternatives.

Harm can be wrought by hunger, emotional abuse, noise, toxic chemicals, radiation, loss of open spaces, or other stresses, as well as by physical violence.

Risk assessment generally focuses on a particular activity, hazardous substance, or project. Alternatives to these activities, substances, or projects are rarely considered or pursued in a meaningful way.

2. *It is not acceptable to harm non-humans when there are reasonable alternatives.*

I am not talking about hunting here, although some people would (Kerasote 1997). I'm talking about the type of harm involved in polluting wildlife with toxic, chlorinated organic compounds (organochlorines) that weaken or kill them.[2]

I am talking about the type of harm involved in building more roads in the last areas where wild wolves (e.g., eastern timber wolves—see figure 1.1) survive. Humans often kill wolves when they see them from roads. Wolves are wary of raising pups near people and cars and roads. In general, wolves will have trouble surviving if there is more than one mile of road in a square mile of land where they are living (USFWS 1992).

While people differ on the value they place on providing health and habitat for non-human beings, such as wolves, butterflies, or fish, this book is based on the assumption that, for their own well-being and that of all living things, human beings need to share the Earth with as broad a diversity of living beings as possible.

Alternatives assessments could technically consider benefits and damages of particular activities only as they would affect human beings. However, many people acknowledge their personal, cultural, spiritual, physical, psychological, and/or economic dependence on a biologically and ecologically diverse world. Many people also recognize a deep connection with the non-human world and feel fully human only when they respect the needs of their non-human co-inhabitants of the Earth. This book is based on such respect.

Again, risk assessments rarely involve serious consideration of viable alternatives.

3. *Nobody is able to define for someone else what damage is "acceptable."*

Alabama State Attorney General Jimmy Evans has explained this well in relation to the chemical called "dioxin" (Kipp 1991):

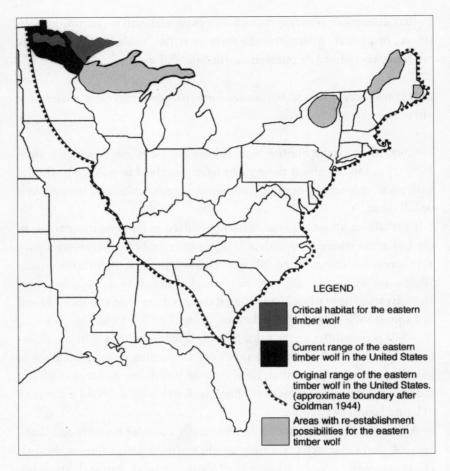

Figure 1.1
Source: USFWS 1992.

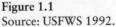

The risk assessment technologies . . . say people will die as a result of dioxin emissions. Then they say that is perfectly acceptable. . . . That is really, really—to me—outrageous and bizarre. It reflects an elitism, a plantation mentality. I think it amounts to a confession. It is very, very simple to me. It is a moral issue. They have said people will die, and we are supposed to accept that. As attorney general of this state, I can't.

Some of the laws and regulations we have developed in the United States require government regulators to determine "acceptable" damage: "acceptable" levels of drinking-water contamination, "acceptable limits of change" on public lands, "acceptable" numbers of cancers that will be caused by

Box 1.1
Dioxin, Dioxin-Like Chemicals, and Organochlorines

The term "dioxin" refers to a group of 75 different chlorinated organic chemicals that contain two "rings" of carbon atoms, two oxygen atoms joining the two rings, and chlorine atoms attached to certain carbon atoms on the carbon rings. Sometimes "dioxin" is used to refer to the most toxic known dioxin, 2,3,7,8-tetra-chlorodibenzo-*p*-dioxin, which is also referred to as 2,3,7,8-TCDD.

Certain (but not all) dioxins cause particular toxic effects in numerous living organisms. These toxic effects include cancer, immune-system damage, imbalance of sexual hormones, nervous-system damage, reproductive failure, birth defects, kidney damage, and skin defects.

"Dioxin-like chemicals" refers to a larger group of chemicals that elicit the same toxic effects as does 2,3,7,8-TCDD in numerous living organisms. These chemicals include certain other dioxins (i.e., those that have a chlorine atom at each of the 2, 3, 7, and 8 positions on the carbon rings, but also have chlorines at other positions on the carbon rings), furans, polychlorinated biphenyls (PCBs), polybrominated biphenyls (PBBs), and polychlorinated naphthalenes (PCNs). The source of nearly all dioxins and dioxin-like chemicals present in the environment is human manufacture, use, and disposal of chlorinated organic (i.e., carbon-containing) compounds.

The dioxin-like chemicals elicit the same toxic effects as 2,3,7,8-TCDD because they attach to the same receptor in cells and lead to production of the same enzymes and toxic chemical reactions in living organisms as 2,3,7,8-TCDD. They do this in weaker fashion, however. Some dioxins and furans, for instance, are one-half or one-thousandth as "potent" as dioxin at eliciting the toxic effects. These "weaker" dioxins and furans, however, may be present as pollutants in the environment in much larger amounts than dioxin. They can thereby cause large amounts of dioxin-like damage.

Chemical compounds that contain carbon are called "organic compounds," and organic compounds that contain chlorine atoms are called "organochlorine compounds" or "chlorinated organic compounds."

source: O'Brien 1990a

some activity like producing or using some toxic, "acceptable daily intakes" of toxic pesticides, and so on. This does not alter the fact that what they have regulated as being "acceptable" may in fact *not* be acceptable to you, to your unborn child, to your community or tribe, or to Florida panthers or wood ducks.

What is acceptable to any person is a matter of personal judgment, but the word is used by risk assessment's promoters as if it were something concrete that could be measured by others, or as if it were something about which everyone must surely agree. This is not accurate. For instance, while a state's Department of Environmental Quality may call some amount of toxic pollution of well water acceptable, a person who actually drinks this well water may not find *any* unnecessary pollution acceptable.

The alternatives-assessment approach is based on the concept that foisting any risk or damage on an unconsenting, unwilling, non-speaking living being or ecosystem must not be labeled acceptable if there are reasonable alternatives to causing that damage, or that much damage.

Decision makers and industry frequently use risk assessment to justify their decisions, as it appears to consist of scientific information (data or assumptions) objectively plugged into mathematical models that supposedly document the minimal, insignificant, or acceptable risk of their proposed activities.

4. Most private behavior has environmental consequences for the public, so it isn't actually private.

If someone stood on his front porch and shot at children walking to or from school, that person would not be able to claim that he can do what he wants on his private property. He may be standing on his own property when he pulls the trigger, but the impact of the bullet will occur somewhere else.

When WTI releases lead (and other toxic metals and chemicals) from its incinerator, it is, in reality, "shooting" at children required to attend and play at the neighborhood school. Lead damages children's brains, kidneys, and blood-forming organs (National Research Council 1993). The National Research Council[3] notes that "there is growing evidence that even very small exposures to lead can produce subtle effects in humans" and admits that the current "standard" of "acceptable" exposure, 10 micrograms of lead per deciliter of blood, may not protect people (National Research Council 1993).[4] The WTI incinerator is permitted to send 9400 pounds of lead into the air of East Liverpool each year (Montague 1992a).

In other words, WTI is a private company that is polluting children's bodies with lead, without the children's knowledge or consent. WTI is impinging on the lives and fates of children whose lives and fates WTI does not own. This behavior of WTI cannot accurately be called "private" behavior, because it contaminates and endangers the public. The public therefore has a right to participate in decision making about whether particular consequences of WTI's activities will be permitted.

Risk assessments tend to involve citizens only as commenters, if at all. In addition, risk assessments are generally difficult for the public to participate in or understand.

5. We humans inevitably cause environmental damage; the only way we will cause the least damage is to consider options for causing the least damage and restoring environmental health when possible.

With our numbers, our technologies, and our consumption, we humans have not proven "safe" for one another, for other animals, or for plants. For instance, few grizzly bears remain in the lower 48 states, where they once were abundant. Less than 50 percent of the wetlands that once flourished in the lower 48 states remain (Noss et al. 1995). The ability of numerous children to remember numbers and words has been damaged because, prior to pregnancy and/or while breast feeding, their mothers ate some fish (about two to three salmon or lake trout meals per month) from the Great Lakes, which have become polluted with toxic chemicals. Among the toxic chemicals these mothers ingested by eating these fish were PCBs,[5] which can damage the brains of developing embryos. The PCBs passed through the mothers' umbilical cords and breast milk to their children when the children were embryos and infants, still developing their brains (Jacobson et al. 1990).

The damage we do, however, can be less dramatic than manufacturing and using highly persistent toxic compounds such as PCBs. Most of us, for instance, have contributed to destruction of the intellectual power of children and other beings by disposing of plastics, toxic household chemicals, or toxic paints in bulging landfills, which occupy land that was once home to wildlife and where water runoff was once clean.

It is not a matter of *whether* we cause damage to each other and other living organisms. Instead, it is a matter of *how much* damage we cause. The major question is whether we and our social institutions (e.g., corporations and legislatures) will approach the world recklessly, causing or permitting as much damage as we can get away with; or carefully, causing or permitting as little damage as possible. If we are going to try to cause as little damage as possible, we (and our social institutions) have to systematically examine options for least damage.

6. It is difficult for many people to think of alternatives to business as usual. Also, it is in the interests of some people (and some corporations) to pretend there are no better ways of behaving, so that they need not change their current behaviors.

Sometimes we are hardly aware that the damage we are causing is unnecessary.

Example: If a father has always killed prairie dogs on his family's cattle ranch because he is convinced that prairie dogs compete with the cattle for forage (edible plants), it may be hard for others in the family to know or acknowledge that they could maintain both prairie dogs and cattle on their ranch, as some ranchers do. Because relatively few prairie-dog towns have been allowed to continue on range land, black-footed ferrets, which eat only prairie dogs, have become among the rarest animals on Earth (Kenworthy 1992). Keeping prairie dogs on a ranch can help with recovery of black-footed ferrets.[6]

Example: We may not be aware that if instead of buying white paper we bought cream-colored paper made without using chlorinated compounds we would be causing less damage to the environment. A paper mill that uses chlorine gas, chlorine dioxide, or hypochlorite to whiten pulp produces and dumps at least half a ton of chlorinated wastes a day into the river or lake from which it draws water. These chlorinated wastes contain as many as 1000 different organochlorines, many of which are known to be toxic to fish, to fish-eating wildlife, and to humans. A pulp mill that does not use any chlorine compound to pulp dumps virtually zero toxic organochlorines into its river or lake (Kroesa 1990).

Part of the reason we allow the environment to be harmed in numerous ways, then, is that we may have only one side of the story. When a corporation sets out to sell products that are made using harmful practices, it tries to convince customers that it has to make the products the way it does, and that the customer really needs and wants those products.

Public consideration of options for environmentally sound behavior destroys the claims that environmentally damaging activities, such as killing prairie dogs or using chlorine compounds to produce paper, are necessary.

By assessing the risks of hazardous activities and concluding that some level of the activity poses no or insignificant risks of damage, the risk-assessment process generally attempts to obviate consideration of serious alternatives to those activities or substances.

7. It is difficult to change many of our habits and behaviors.

It is hard for an individual, a company, or a government agency to change habits. Habits include thinking in certain ways, following certain bureaucratic processes, selling to certain markets, using certain equipment or manufacturing processes, or even listening only to certain people. We tend to be creatures of habit. As the writer Terry Tempest Williams said to me when we were temporarily "lost" in the snow on a small butte where I had hiked many times, "once you step out of your ordinary track, you don't know where you are at all."

Change is often scary, economically or socially disruptive (at least temporarily), anxiety-producing, or technically difficult. All strategies for improving human approaches to the environment ought to take into account the economic, political, structural,[7] psychological, and emotional bases of resistance to change.

For the most part, risk assessments are used to justify business as usual and to marginalize calls for change.

8. We have no choice but to gain practice in changing our environmentally bad habits everywhere.

Most people are aware of numerous environmental problems, many of which we suspect will only get worse in the future, such as population

growth and excessive consumption, loss of wildlife habitat, water short-
ages, and global warming. We know we have to deal with these issues, but
none can be resolved until we change certain ways we behave. Therefore,
we have to change how we behave.

For the most part, risk assessments are used to claim that change is not
necessary.

9. One of the most essential prerequisites for political change is under-standing that there are alternatives.

Democratic processes for change have always been triggered when some
people have become convinced that undesirable conditions are not
inevitable and that a better approach is possible. These democratic processes for change include writing laws, passing initiatives, exerting community
pressure on local businesses, litigation, boycotts, market pressures, incentives, and protests. Once people learn about attractive alternatives, there
are many democratic routes to their implementation.

We need to establish public procedures that ensure that alternatives for
treating the world and each other better will be publicly discussed and considered. We need to publicly discuss both the benefits and the drawbacks of
each alternative.

By claiming on a supposedly objective basis that current activities pose
little if any risk of damage, risk assessors generally downplay the need to
even consider alternatives.

10. Changes in the damaging behaviors and habits of other people (and corporations) must be accomplished through political action.

Ultimately, we need to implement environmentally protective and socially
respectful alternatives through democratic processes that communities,
states, nations, and the international community have created for enabling
people to protect themselves, the public health, and the environment.

Changes do not come easily when powerful entities benefit from the status quo. Change is accomplished through political action, and political
action is the responsibility of citizens in a democracy if they are truly to

govern themselves. Citizens include scientists, parents, bureaucrats, teachers, physicians, judges, workers, and children.

As Frederick Douglass wrote in 1857:

If there is no struggle, there is no progress. Those who profess to favor freedom, and yet deprecate agitation, are men who want crops without plowing up the ground, they want rain without thunder and lightning. They want the ocean without the awful roar of its many waters.

This struggle may be a moral one, or it may be a physical one, and it may be both moral and physical, but it must be a struggle. Power concedes nothing without a demand. It never did and it never will.

Risk assessment is an extremely flexible and powerful tool for dispelling calls for change.

2

How Does Risk Assessment Actually Work?

In theory, risk assessment is an objective, science-based process. In reality, risk assessment involves choices among numerous "guesses" and estimates. Politics, money, and power affect those choices.

The Theory of Risk Assessment

Hundreds of conferences and workshops are held each year in the United States, either to "explain" to an audience of interested citizens how risk assessment works or to teach people how to conduct risk assessments. These meetings are held because decisions in so many areas—from using pesticides to siting waste facilities, building energy-generating plants, mining for coal, and dozens of other activities—include risk assessment as one step in the decision-making process.

If you were to attend one of these conferences (say, one on the use of a pesticide), the presenter would begin by describing what risk assessment is and would then explain that the risk inherent in the pesticide depends on two elements: *hazard* and *exposure*. Hazard means the toxicity of the material; that is, its capability to cause particular kinds of damage. This evaluation is usually based on results from laboratory animal experiments; sometimes there are observations from human experience. Exposure is the amount of pesticide (the "dose") that is actually delivered to the living organism whose harm is being measured, usually a human. The amount of exposure will vary with the way the pesticide reaches the organism—for example, through the skin, or through inhalation or ingestion. Estimates may be made of the degree to which the chemical will actually reach the organs where it would cause damage.

Box 2.1
Some Uses of Risk Assessment in the United States[1]

Food
Pesticide registration
Allowable pesticide residues
Allowable hormone residues
Allowable drug residues

Water
Allowable withdrawals of water from rivers and aquifers
Sediment load limits
Allowable drinking-water contamination (e.g., maximum contaminant levels)
Allowable groundwater contamination
Permits to discharge toxic substances to rivers
Roadside spraying

Air
Permits to discharge particulates and toxic chemicals to ambient air
Aerial pesticide applications
Indoor air limits for toxic chemicals
Auto emissions limits
Burning/incineration permits

Land, oceans
Livestock grazing allotments in arid lands
Logging permits
Ecological risk analysis
Permitted takings of endangered species
Fishing and hunting quotas
Hatchery releases
Road construction

Workplace
Limits for toxic chemicals in workplace (e.g., threshold limit values)
Worker re-entry intervals after pesticide applications in fields
Radiation limits in nuclear workplaces

Cleanups
Cleanup standards for leaks, spills, Superfund sites, nuclear sites

Prioritizing funds and action in the face of multiple environmental problems
Comparative risk assessment

source: Mary O'Brien

_____

1. Risk assessment is often "informal," using few numerical calculations or mathematical models.

Box 2.2
Hazard, Exposure, and Risk in Quantitative Risk Assessments

All risk assessments are based on combining the hazard of an activity or a substance and the exposure of humans or other species to the hazardous activity or substance in order to estimate risk of damage by that activity or substance to humans or other species.

Hazard: Damage That Might Be Caused by an Activity or a Substance

Hazards describe the damage that can be caused by something such as a road, global warming, radiation, a chemical weapon, or a desert motorcycle race. The hazards considered are usually physical adverse outcomes such as death, cancer, nerve damage, reduced photosynthetic activity of plants, altered maternal behavior (e.g., of a nesting bird), or loss of hearing.

Hazards posed by an activity or a substance are often calculated by experimenting on animals, plants, model ecosystems (e.g., one small lake), or other experimental organisms.

Alternatively, hazards can be estimated by measuring what happened when people or other species were exposed to some estimated or measured amount or intensity of a substance or activity in a real-life situation. For instance, frogs' eggs may have reduced hatching success when they were deposited in a lake exposed to increased ultraviolet light. Some firefighters may have gotten sick when they were exposed to a spilled industrial chemical.

Such experiments or observations usually seek to determine some amount of exposure to the hazardous activity or substance that does not cause adverse effects.

Exposure: Amount or Intensity (Dose) of the Activity or Substance That Some Humans or Other Species Might Experience

Risk assessors then estimate the degree to which someone or something (e.g., a frog, a human, a community, a desert lizard) would likely be "exposed" to the hazardous activity or substance. They estimate the "dose" of the hazard that someone (usually an "average" member of some group) will experience.

Exposure can be estimated from controlled experiments, measurements taken under work place or field conditions, or accidental exposures. Often, exposure is simply estimated from closely or distantly related situations, or, at worst, simply chosen to give predetermined risk outcomes.

Many factors can be considered when guessing what the exposure might be, such as amount of skin exposed to a chemical in the air, percent of a chemical that will be absorbed into the body through the skin, percent of absorbed chemical that will reach the vulnerable organ or tissue, amount of food consumed, direction and speed of the wind, rate at which rainfall replenishes an underground aquifer, numbers of hours or days an organism is exposed, body weight, and so forth.

Inevitably, many factors are ignored, either because they have never been examined, because their inclusion would make the quantitative risk model too complicated, or because their inclusion would make the exposure seem too dangerous. (See boxes 2.4 and 2.5.)

Risk: Will Someone Be Damaged? How Much?

On the basis of their estimate of the ability of an activity or substance to cause a particular kind of damage (hazard), and their estimate of the degree to which a certain organism will be exposed to that hazard (exposure), risk assessors finally calculate the likelihood of organism experiencing that damage (risk).

source: Mary O'Brien

Box 2.3
Enormously Simplified Models of Quantitative Risk Assessments

Non-Cancer Effect

The risk of a non-cancer effect (e.g., birth defects) after exposure to a toxic chemical is calculated by dividing *hazard* (i.e., highest NOEL[1]) by *exposure*, where *risk* means how close humans or other animals will come to experiencing adverse effects observed in non-human laboratory animals given the exposure expected in the situation for which the risk is being calculated.

Example
Some governmental bodies (e.g., Canada, the World Health Organization) have indicated that a dose of 10 pg/kg dioxin is "tolerable" in terms of reproductive and developmental effects (Webster and Commoner 1994).[2] In this example, the hazard is a NOEL of 1 ng/kg/day dioxin. This is based on a multigenerational rat fertility experiment that showed exposure-related effects on fertility, an increased time between first cohabitation and delivery, and a decrease in litter size (Murray et al. 1979). A safety factor of 100 was given to this, on the assumption that humans might be 10 times as sensitive as a rat and that some humans might be ten times more sensitive than the average human. The exposure is an average human daily dose of 1–3 pg/kg toxic equivalents of dioxin. Thus, the risk is calculated as follows[3]:

(0.01 ng/kg/day)/(1–3 pg/kg/day).

This says that humans are estimated to be 3–10 times below the exposure that might cause reproductive difficulties.[4]

Cancer Risk

Cancer risk (i.e., how many people per million are likely to contract cancer as a result of the exposure being examined) after exposure to a cancer-causing chemical is calculated as the product of hazard (i.e., cancer potency of chemical) and exposure (i.e., dose).

Example

The U.S. Environmental Protection Agency (USEPA 1994e) estimated the cancer risk to a private applicator spraying oryzalin, a 2,6-dinitroaniline herbicide produced by DowElanco. Oryzalin is used to kill annual grasses and broadleaf weeds on berries, vine and orchard crops, Christmas tree plantations, commercial/industrial and recreation area lawns, golf course turf, residential lawns and turf, ornamental and/or shade trees, non-agricultural rights of way/fencerows, non-agricultural uncultivated and industrial areas, power stations, paths/patios and paved areas. Oryzalin is also used to control herbaceous plants, woody shrubs, and vines. Oryzalin has been found to cause mammary gland tumors (i.e., breast cancer) in female rats and skin and thyroid tumors in male and female rats. EPA considers oryzalin a Group C carcinogen (i.e., possible human carcinogen). Cancer potency = 0.13 mg/kg/day. To calculate the cancer risk to a "private applicator," the following assumptions are made: The applicator will apply oryzalin one time, one year in her/his life. The applicator will use a low-pressure hand wand to apply the oryzalin. The applicator will wear long pants, a long-sleeved shirt, and no gloves. The applicator's dermal exposure will be 52 mg per pound oryzalin applied. The applicator will apply 0.094 pound oryzalin per one thousand square feet. The applicator will apply oryzalin to two acres. The private applicator's lifetime average daily dose = (total daily dose × work days per year × 35 years)/70 years. Therefore, the applicator's lifetime average daily dermal exposure will be 2.0×10^{-4} (i.e., 0.0002) mg/kg/day.

Hazard (0.13 mg/kg/day) × Exposure (0.0002 mg/kg/day) = Risk (2.6×10^{-5}).

In other words, from spraying oryzalin one day in one year, a private applicator will have an excess cancer risk of 2.6×10^{-5}, or 2.6 cancers/100,000 people, which is 26 times more than one cancer in a million.

Ecological Risks

Ecological risks (e.g., some risks posed by elevated ozone in a region of varying forest types) are also calculated as Hazard × Exposure.

Examples

One ecological risk analysis (Graham 1991) estimated consequences of regionally increased ozone [Hazard] in the lower portions of the atmosphere, which causes physiological stress in coniferous pine trees [Hazard], which could lead to tree death from vulnerability to a pine bark beetle infestation [Hazard]. Using some data (e.g., from bark beetle infestations in ozone-stressed ponderosa pine trees in the San Bernardino Mountains of California), computerized risk analysis focused on estimating consequences [Risk] of reduction of pine tree cover from such infestations in the Adirondack region of New York. Predicted consequences [Risks] included a drop in acidity in lakes near forests and alteration of wildlife habitat at the edge of forests. The model predicted little change in the interior areas of such forests.

Another ecological risk analysis (Müller et al. 1997) predicts that the severe ozone depletion [Hazard] that was observed in the Arctic during the winter of 1995 and 1996 will recur, because cooling of the stratosphere (the layer of atmosphere in which the Earth's protective ozone layer lies) in the Arctic and human-increased concentrations of chlorine chemicals in the stratosphere [Exposure] are expected to persist for several decades. Other risk assessments of these conditions (summarized in Makhijani and Gurney 1992) have predicted various percent increases in skin cancers (melanoma and non-melanoma), cataracts of the eye, reduction in soybean yield and reduction in shrimp production.

source: Mary O'Brien

---

1. The No Observed (Adverse) Effect Level (NOEL) is the highest experimental dose (e.g., 6 mg pesticide per kg body weight of the test animal) that does not produce any of the adverse effects (e.g., liver damage) that are being observed in that animal.

2. One picogram/kilogram is equivalent to one part per quadrillion, or one second in 32 million years.

3. 0.01 ng/kg/day is equal to 10 picograms/kilogram/day. This is the hazard, with a 100-fold safety factor.

4. This risk assessment, like most, has been questioned (Webster and Commoner 1994). For instance, rhesus monkey studies have shown a Lowest Observable Adverse Effect Level (LOAEL; no NOEL was found) of 0.64 ng/kg/day chronic exposure to rhesus monkey to lead to prenatal mortality (Bowman et al. 1989a) and 0.13 ng/kg/day to result in decreased learning among offspring (Bowman et al. 1989b).

Based on hazard and exposure, the risk assessor would estimate the resulting risk that exposure to this pesticide poses. For instance, the risk assessor's formulas might predict that a commercial sprayer of a potent carcinogenic pesticide will have twice the risk of getting cancer as someone not exposed to the pesticide. Or the risk assessor might predict that a particular spray program will expose robins to 3 times less pesticide than would cause birth defects.

A pesticide risk assessment could examine effects other than direct toxic effects. An assessment of risks of an insecticide, for instance, might discuss the possibility that, if the intended insect pest is killed, a "secondary" insect pest may then become more numerous.

Of course, the risks of any activity or substance can be assessed. For example, a risk assessment might examine the risks to eagles and hawks of getting caught in the blades of wind-power generators, the risks to villagers' eardrums from noise in a neighboring rock quarry, or the risks to cattle of losing their calves to brucellosis, which is caused by a bacterium carried by some bison and elk in Yellowstone National Park (Keiter 1997).

In this presentation on the basics of risk assessment, however, we are getting only the theory of risk assessment, not the reality. Let us look at some of the realities behind risk assessment. Again, we will take the example of a risk assessment of a pesticide, but the realities of pesticide risk assessments are similar for risk assessments of all kinds of activities and substances.

First, a risk assessor cannot know the hazards of the pesticide. (See boxes 2.4 and 2.5.) For instance, she or he will not know all the toxic effects of the pesticide or of the chemicals into which the pesticide transforms out in the environment. Further, the risk assessor cannot estimate all the real-world toxic effects to which the pesticide will contribute, because the assessor doesn't know the victim's particular sensitivity to particular pesticides or the victim's exposure to numerous other toxic chemicals, radiation, malnutrition, drugs, particulates, and other stresses. These other exposures will influence the pesticide's additional effects.

Second, the risk assessor will have to make numerous "guesses" or assumptions to estimate how much some hypothetical organism will be exposed to the pesticide in the real world—for example,

• how much of a given item of food the "average" person eats if the pesticide is allowed on that food

Box 2.4
Hazard Elements Considered and Those Ignored in a Quantitative Risk Analysis

Considered
Hazards noted only in laboratory experiments using controlled dose of pesticide and length of exposure
Hazards only of the revealed ingredient(s) of a pesticide formulation
Hazards to non-human laboratory animals of homogeneous genetic strain
Laboratory animals exposed to one test chemical
Specific adverse effects noted for the specific test
Effects noted in required tests whether or not they have been completed adequately

Ignored
Hazards experienced by humans and non-humans in the field following exposure to undetermined dose or exposure of undetermined duration (These are the effects experienced in reported and unreported incidents and accidents, human epidemiological studies, fish kills, etc.)
Hazards of the full formulation, including secret ingredients (solvents, preservatives, petroleum distillates, emulsifiers, etc.) alone and in combination with the revealed ingredients; metabolites, degradation products, and carriers
Hazards to humans and non-humans varying greatly in genetic characteristics, age, sensitivity, and health condition
Additive, cumulative, and synergistic hazards of daily exposure to other toxic chemicals (e.g., pesticides in food, air contamination) as well as the pesticide in question
All possible effects, including those not detected in laboratory tests (e.g., headaches, joint pain, fatigue, emotional perturbation)
Effects that might have been noted if required tests had been completed or performed adequately, or if tests not currently required (e.g., nerve damage, immune suppression) had been performed

source: O'Brien 1988

- what size and age the exposed person is[1]
- how fast and in what directions the wind blows on an average day if the pesticide will be released into the air
- how much protection is offered by generic clothing or by well-functioning "protective" equipment if the person being exposed to the pesticide is a worker
- how often a truck carrying the pesticide can be expected to have an accident and spill the pesticide.
- how sensitive the exposed person is to the particular pesticide.[2]

Box 2.5
Exposure Elements Considered and Those Ignored in a Quantitative Risk Analysis

Considered
Estimated exposure, generally based on some measurement, in some closely or distantly related situation exposure to the revealed ingredient(s), which is often a very small proportion of the volume of the entire pesticide formulation

Generalized estimation of drift, water contamination, food residues, skin absorption

Exposure to adult body (e.g., 70 kg)

Exposure only to this pesticide for the planned program, the risk of which is being estimated

Ignored
Actual exposure in various distinct situations

Exposure to the full formulation, including secret ingredients, carriers, contaminants, and metabolic and degradation products

Uneven field concentrations of pesticides in fog, rain, air currents, volatilization, and through leaky backpack sprayers, used work clothes, sensitive areas of the body

Exposure to infants, small children

Exposure to this and other pesticides in other past, current, and future programs of pesticide use

source: O'Brien 1988

Some people argue that the transparency of risk assessment—that is, the laying out of all numbers in the formulas and models, is a strength of the process. They argue that this transparency opens the risk assessment to public criticism and debate, which improve it. Theoretically, this is true. However, in reality there are many estimations and selected numbers in most risk assessments, and the means available to significantly change an official risk assessment using alternative numbers are considerably unequal. Corporations can and do plug in different estimated numbers so as to reach different conclusions, because they often have millions of dollars riding on the outcome of a risk assessment of one of their products or activities. Risk-assessment companies can do this for their contractors who can pay them to do it, and some government agencies have the means to do it too. However, a single independent scientist or knowledgeable citizen will have difficulty arguing with the numbers that a risk assessor plugs into a hazard or exposure estimate. Who is to say that the numbers such an individual would put into the exposure estimate are more accurate? Who is to say that the research an individual cite is any more valid than the research the corporation has produced, or that the risk assessor cites? If no research is available for a particular number that has been placed into the hazard or exposure model, how likely is it that a citizen's report of what that number should be will be adopted if it shows more risk than the "expert" risk assessor's estimated number or the government agency's estimated number?

In other words, the theoretical "transparency" of risk assessment does not necessarily translate into a level playing field for altering a risk assessment which concludes that a certain hazardous activity is "safe," or of "insignificant risk," or "acceptable."

Many people know of William Ruckelshaus's famous description of risk assessment (1984): "We should remember that risk assessment data can be like the captured spy: If you torture it long enough, it will tell you anything you want to know."[3] When he said this, Ruckelshaus was the Administrator of the U.S. Environmental Protection Agency, and he was promoting the establishment and use of risk-benefit analysis within that agency. Previously, Ruckelshaus had said: "In assessing a suspected carcinogen, for example, there are uncertainties at every point where an assumption must be made: in calculating exposure; in extrapolating from high doses where we have seen an effect to the low doses typical of environmental pollution; what we may expect when humans are subjected to much lower doses of a substance that, when given high doses, caused tumors in laboratory animals; and finally, in the very mechanisms by which we suppose the disease to work." In addition, however, Ruckelshaus argued that risk assessment ought to be done. According to him, if our democratic society is to remain "grounded in a high-technology industrial civilization," we need to move from trying to ensure a "margin of safety" for the public to making decisions on the basis of risk-benefit analysis. "We must assume," he noted, "that life now takes place in a minefield of risks from hundreds, perhaps thousands, of substances." In other words, Ruckelshaus saw the industrial "need" to expose people to toxic substances, saw the flexibility of the risk-assessment approach for calculating what toxic risks will be allowed, and wanted to establish this risk-assessment approach within the EPA. (See, e.g., Merrell 1983.)

However, the EPA, other agencies, or corporations cannot easily be held accountable for whether the conclusions of a risk assessment reflect reality, because the conclusions of a risk assessment are so mutable.

The Theoretical Separation of Risk Assessment and Risk Management

In our pesticide risk-assessment workshop, the presenter will almost certainly explain that risk assessment (i.e., determining hazard, exposure, and risk) is separate from risk management. Risk management consists of the social, economic, and political decisions that will be made about who will bear what amount of exposure to the pesticide. In other words, the presenter will tell us that there is something scientific and objective (called *risk assessment*) and then there is something political and value-laden (called *risk management*).

In reality, however, most risk assessments are prepared when permission is sought by a business, an agency, or a corporation to initiate or continue a hazardous activity or to use a poison. That is, a risk-management decision that will have consequences for a business or an agency is already on the horizon. The risk assessor is generally hired by private industry or by the government to do a risk assessment of a value-laden and sometimes highly controversial situation. Most risk assessors do not stay clear of risk-management considerations during the process of estimating risk. Since there is a wide choice of which numbers will be plugged into a risk assessment, and since no one usually knows for sure what is the "right" number to use, the pressure on a risk assessor to use numbers that will fulfill the wishes of the company or agency by which she or he is employed becomes tremendous. The bottom line in most (if not all) risk assessments is that if someone wants to continue some activity or to get a permit or approval for some activity, and if the outcome of the risk assessment will get in the way of that activity, there will be pressure to use optimistic numbers in the risk assessment.

On the other hand, if a citizens' organization is able to secure the services of a risk assessor, the organization will generally work with someone who tends to assume that if something dangerous might happen then that assumption should be in the risk assessment.

What this means is that risk assessments are generally not separate from risk management.

Risk Assessment and Risk Management in Practice Rather Than in Theory

We have heard the workshop presenter's picture of the theory of risk assessment and risk management. Let us now look at some examples of how risk assessment and risk management actually work.

1. How much workplace exposure to that toxic chemical is safe?

Threshold Limit Values (TLVs) are the average concentrations in air for an 8-hour work day, in a 40-hour work week, to which "nearly all workers may be repeatedly exposed, day after day, without adverse effect" (Castleman and Ziem 1989). Since 1946, TLVs have been set by the Committee on Threshold Limits of the American Conference of Governmental Industrial Hygienists (ACGIH). The ACGIH is a private organization composed of industrial hygienists from state and local governments, plus academics and industry consultants. The U.S. Occupational Safety and Health Administration (OSHA) has adopted most of the TLVs prepared by this private organization as official, enforceable, permissible exposure limits in workplaces in the United States. Many of the standards now in use were developed in the late 1960s.

Castleman and Ziem (1988), reviewing almost 600 TLVs, reported that at least 104 of them relied importantly or exclusively on unpublished information from corporations. This corporate information was often provided to the TLV committee by industry scientists whose corporate employers would have to restrict their exposure of workers to the toxic chemicals if the TLVs were low. In other words, it was important to the corporations that the TLVs not restrict their use of the toxic chemicals. A higher TLV would mean that more exposure of their workers to the toxic chemical would be legal.

Roach and Rappaport (1990), comparing TLVs to the scientific reports that members of the Committee on Threshold Limits said they had used to set the TLVs, found that in numerous cases the TLVs had been higher than levels that the scientific reports had shown to cause adverse effects in humans. The adverse effects ranged from eye and nose irritation to disease and permanent bodily changes.

In 1976, the ACGIH said this of the ethyl ether TLV: "Regular exposure at this concentration (400 ppm, the TLV) should cause no demonstrable injury to health nor produce irritation or signs of narcosis among workers."[4] One study ACGIH cited, however, indicated that "complaints of nasal irritation began at 200 ppm in the majority of 10 volunteers."

The isopropyl acetate TLV was explained by ACGIH as follows: "The limit, 250 ppm . . . is considered adequate to prevent significant irritation of the eyes and respiratory passages." The ACGIH cited an isopropyl acetate study, however, that reported the following: "We found that at 200 ppm, the majority of . . . twelve subjects of both sexes . . . experienced some degree of eye irritation."

The ACGIH said that for fluoride as F (fluorine) "the limit, 2.5 mg/m^3, is sufficiently low to prevent irritative effects and to protect against disabling bone changes." The ACGIH was aware, however, that at a factory where the concentration of fluorides ranged from 0.14 to 3.13 mg/m^3 "radiological [x-ray] examination revealed signs of osteosclerosis [abnormal hardening of bone] in 48 of 189 workers." What this means is that some of the risk assessments of worker exposure to particular chemicals simply "overlooked" some of the data, with the result that the industries using these chemicals would not have to reduce or eliminate the release of these chemicals in their workplace.

These examples show that numbers representing accepted TLVs that are plugged into risk assessments bear the danger of being biased by political or economic factors. They look like scientific numbers, but they don't necessarily correlate with reality as observed and documented.

2. How much fish can we harvest?

Ludwig, Hilborn, and Walters (1993), three fisheries researchers, note that "there is remarkable consistency in the history of resource exploitation"—namely, that resources such as fisheries, forests, soil life, and grasslands are "inevitably overexploited, often to the point of collapse or extinction." These authors describe four conditions that lead to optimistic risk assessments regarding exploitation of natural resources:

• Wealthy resource exploiters or the prospect of getting wealthy from overexploitation of a resource generate political and social power that is used to promote unlimited exploitation of the resource (e.g., by hiring "optimistic" scientists). A fishing corporation, for instance, will hire fisheries scientists who the corporation knows are likely to estimate that there are more than enough of the fish available in the ocean.

• Each new resource problem involves learning about a new system (e.g., what constitutes overfishing of a particular fish species), and the scientists generally don't yet have studies showing the comparative effects of exploitation and non-exploitation of that resource. If some scientists, for instance, say that they are worried that sardine populations may "crash" like certain other fish species did if they are fished as heavily as proposed, the fishing corporation will likely say "You haven't given evidence that sardines will crash."

• The natural resource system is complex. Therefore, "reductionist" (i.e., simplified) models and examples are not sufficient to estimate the effects of exploitation on the system. For example, a fishing risk-assessment formula that concentrates only on sardines or only on one stress (such as commercial fishing) on sardines is not realistic enough to predict how many sardines can be sustainably exploited (fished). This is because other things may be happening in the ocean. For instance, other fish species may depend on sardines for their own food, so depleting sardines may cause starvation of their predators. Sardine populations may already be weakened by pollution or temperature changes, or their own food base may have been depleted so the resiliency of their response to being fished may be diminished. Because multiple species often depend on each other in unsuspected ways, and because they are often exposed to multiple stresses, simple risk-assessment formulas often gravely underestimate the consequences of particular human exploitive activities.

• Since the condition of a resource differs in different years (e.g., drought years versus years of high rainfall), overexploitation gets hidden. For instance, the fishing corporation might say "Well, this year, the warm ocean current called El Niño is the cause of failure in the sardine fishery, but the fishery will rebound." Ludwig et al. (1993) note that the idea of maximum sustained yield (MSY) of a fishery (i.e., the largest possible number of fish that can be regularly harvested without causing the fish population to collapse) is now widely recognized as an "unfortunate" concept. Few fisheries exhibit steady abundance, and the MSY constitutes overexploitation during many years. For example, when a fishery experiences a number of good years, businesses invest in new fishing vessels or processing plants. When the fishery returns to normal or below normal, the industry appeals to the government for permission to continue fishing at high levels, because the investments and many jobs are at stake. This concept is clearly applicable to many activities that affect habitat, such as logging, livestock grazing, and mushroom picking. The idea that scientists (e.g., risk assessors) will agree on what constitutes overexploitation is unfounded. When the California sardine was first being exploited, scientists at the California Division of Fish and Game warned that fishing of sardines could not continue to increase without overexploiting the resource, and they recommended an annual sardine quota, or limit, on the amount of the fish that could be taken. However, according to Ludwig et al., the fishing industry "was able to identify scientists [i.e., risk assessors] who would state that it was virtually impossible to overfish a pelagic [oceanic] species." The sardine fishery eventually collapsed, but even today scientists do not agree as to why. Finally, Ludwig et al. note that the type of experiments that are necessary to demonstrate whether a resource is being overexploited involve short-term losses for the industry, because such experiments require some part of the resource to be left free from exploitation, or to be approached with a much lighter hand. An experiment that significantly reduces fishing for sardines in a particular area for 3 years, for instance, might show how quickly sardines can rebound from overfishing. This experiment would mean that fewer fish could be caught in that area during those years, however. Further, the experiment might show that fishing rates for sardines should be lowered everywhere. Although this might ultimately help the sardine fisheries to survive indefinitely, it would cut into short-term profits. Therefore, the experiment would likely be resisted. Ludwig et al. describe some economic, real-world complexities that lead to risk assessments that show "no problem" with human exploitation—risk assessments that deny overexploitation. Their conclusions about exploitation of fisheries clearly apply to similar industries, such as forestry (e.g., how many trees can be logged in a forest area) and agriculture (e.g., how much water can be taken from a stream for crop irrigation), and to suburban development (e.g., how many acres of wetlands can be drained). In fact, they apply to risk assessment of all human activities in which economic or other social impacts are feared.

3. How safe is it to live next door to a dioxin incinerator?

Dioxin incinerators are used to destroy dioxin that has become concentrated somewhere and which therefore poses an extreme hazard. For instance, soil from around a pesticide factory that has been highly contaminated with dioxin or from an area where dioxin-contaminated oil has been spread to keep dust down may be sent to a dioxin incinerator in order to "destroy" the dioxin.

Federal law requires that an incinerator licensed to burn dioxin be able to achieve a "dioxin destruction and removal efficiency" (DRE) of 99.9999 percent (referred to as "six nines"). Such a high rate of destruction is required because dioxin (2,3,7,8-TCDD) is extraordinarily toxic (Montague 1994c,d). The EPA notes in its reassessment of the dangers posed by dioxin

(USEPA 1994a) that "some of the effects of dioxin and related compounds have been observed in laboratory animals and humans *at or near* levels to which people in the general population are exposed" (emphasis added).

The DRE is calculated by subtracting the incinerator's output (release) of dioxin from the input of dioxin (i.e., the amount of dioxin that was fed into the incinerator) and dividing this all by the input. There is an error associated with this simple calculation, however, because the amounts of dioxin are normally extremely small relative to the amounts of soil or other chemicals present in the material being fed into the incinerator. This makes it difficult to accurately measure whether the incinerator is actually destroying the dioxin to six nines. Therefore, dioxin incinerator regulations allow an incinerator operator to demonstrate that the incinerator will destroy the dioxin to six nines by burning to six nines a "surrogate" chemical that is more difficult to burn than dioxin.

In 1991, when the incinerator company Vertac Site Contractors (VSC) applied for a permit to burn dioxin-contaminated herbicide wastes of the Vertac Chemical Company in Jacksonville, Arkansas, it was required to undertake a trial burn demonstrating the incinerator's DRE. VSC was planning to demonstrate the DRE on a surrogate chemical, hexachlorobenzene (HCB), and to burn 2,4-D production wastes[5] and other wastes at the same time. Both VSC and the EPA assumed that the wastes contained no dioxins. However, when Pat Costner (then the research director of the U.S. Toxics Campaign of Greenpeace USA) obtained and read the information on what chemicals went into the incinerator and what chemicals left the incinerator in the stack gases, she found otherwise (Costner 1992). The wastes that entered the incinerator (i.e., the input) contained 2,3,7,8-TCDD, and so did the stack gases leaving the incinerator (i.e., the output). Costner calculated the DRE and showed that the incinerator burned only 99.96 percent of the 2,3,7,8-TCDD. That is, the incinerator burned to three nines, not six nines. This means that the incinerator was going to release much more dioxin into the community of Jacksonville than the EPA had predicted publicly. Neither the EPA nor VSC reported the DRE of the incinerator for 2,3,7,8-TCDD.

When Costner publicly released her calculations, the EPA first claimed that Greenpeace was mistaken. Later, upon examining the data and doing their own calculations, the EPA admitted that Costner was correct, and, furthermore, indicated that they had known for 7 years that dioxin incinerators likely would not destroy dioxin to six nines (Olexsey 1985).

In subsequent litigation, Jacksonville citizens and Greenpeace sought to halt the burning of dioxin-contaminated wastes by the incinerator (which received its permit on the basis of the 1991 trial burn, even though the trial burn showed that the incinerator was not destroying dioxin to six nines). The EPA contended in court that it isn't required to burn dioxin to six nines; that it only has to burn the surrogate chemical, which is supposedly harder to burn than 2,3,7,8-TCDD, to six nines. (See box 2.6.)

The federal district court judge ruled in favor of the citizens and against the EPA because the incinerator was not meeting the DRE of six nines (*Arkansas Peace Center, et al., v. Arkansas Department of Pollution Control and Ecology, et al.,* No. LR-C-92-684, Order (E.D. Ark. Oct. 30, 1992)). When the defendants (including the EPA) appealed this decision, the appeals court ruled in favor of the EPA, allowing them to continue to incinerate dioxin. The appeals court said that citizens cannot sue while a Superfund site[6] such as Vertac Chemical Company is being "cleaned up" (in this case, by incinerating dioxin) (*Arkansas Peace Center, et al., v. Arkansas Department of Pollution Control and Ecology.* 999 F.2d 1212); they can sue only after the cleanup is completed. Moreover, the appeals court indicated that, if they had needed to rule on whether the EPA really has to destroy dioxin to six nines, they would have ruled that the EPA does not have to do so; it only has to destroy to six nines the surrogate chemical, even if the EPA is failing to destroy the dioxin to six nines.

This example illustrates three critical things that often are true of risk assessments in the real world:

Risk-assessment assumptions may not be realistic. In the above example, it was not realistic to assume that a dioxin incinerator will be releasing no more than 0.0001 percent of the dioxin it is "fed." A risk assessment of the dangers posed to nearby community residents by a dioxin incinerator would underestimate the dangers if it assumed 99.9999 percent of the dioxin was being destroyed in the incinerator.

When risk-assessment assumptions are painstakingly debunked, the risk assessors may simply change their argument. When Pat Costner showed that the EPA's assumption of a DRE of six nines was not accurate, the EPA simply responded that they had known that already, and that they aren't legally required to ensure that dioxin is destroyed to six nines. When one

Box 2.6
The EPA: It's the Prediction of Dioxin Destruction That Matters, Not Reality

The following are excerpts from the transcript of a court trial (Arkansas Peace Center v. Vertac Site Contractors, Case No. LR-C-92-684, Eastern District of Arkansas, Western Division, Little Rock, Arkansas, February 12, 1993. Transcript of Hearing before Stephen M. Reasoner, United States District Judge) in which citizens (plaintiffs) sought to halt incineration of dioxin-contaminated wastes in their community. The plaintiffs showed the court that the incinerator was not destroying small amounts of dioxin as completely as had been predicted (i.e., 99.9999 percent, termed "six nines"), and therefore, more dioxin than predicted in risk assessments was being discharged out of the incinerator.

In this transcript, Mr. Spritzer, representing the U.S. Environmental Protection Agency (the defendant), is explaining to Judge Reasoner (i.e., "The Court") that it is sufficient for the incinerator to destroy most of the surrogate chemicals (termed principal organic hazardous constituents, or POHCs) that supposedly are harder to burn than dioxin, even if the incinerator is in fact failing to destroy the dioxin.

page 70 at 16:

The court: And then under your position there is never any subsequent review of what's actually going on? For instance, the question I have asked before. If in fact I allow you to go ahead with your burn, and the data received shows that you are only destroying 50 percent or less of the dioxin, since you demonstrated your ability to destroy the POHCs at six nines, there's nothing I or anybody can do to stop you if you're determined to go ahead?

Mr. Spritzer: Well, we're getting into a somewhat broader issue.

The court: No. I want you to answer that question for me, because I've asked it and it's never been answered. Is that what you contend the regulation provides? Once having demonstrated six nines on the POHCs, it makes no difference what you actually achieve on dioxin?

Mr. Spritzer: For purposes of demonstrating compliance with that regulation, no, it makes no difference, that's correct.

page 82 at 21:

The court: Now, I would assume then that the [U.S. Environmental Protection] Agency's position would be that if the concentrations got low enough as they approached zero, you wouldn't have to be achieving destruction of dioxin at all?

Mr. Spritzer: Well, that's a kind of theoretical question.

> *The court:* Well, I think you and I actually agreed under the Agency's inter-
> pretation, even with very high concentrations of dioxin, you do not have to
> demonstrate any destruction of it at all, as long as you had demonstrated
> your six nines on the POHC.
> *Mr. Spritzer:* That's correct. But at those higher concentrations of dioxin—
> *The court:* We all would hope you would stop it before it came to that
> point, but there's nothing, under your interpretation, a federal court could
> do to stop the burn if it were spewing out dioxin, if you had met your tests?
> *Mr. Spritzer:* That's correct.

court found that argument invalid, a higher court said the citizens couldn't
sue the EPA until the dioxin "cleanup" in Jacksonville was completed.

*Risk assessments can be created so as to benefit not the exposed victims but
the industry or government agency involved.* In this example, for instance,
the EPA took the "side" of defending incineration of dioxin rather than the
side of precaution for the residents of Jacksonville. When this happens, risk
assessment itself is a "management" tool, a policy tool.

4. Risk assessment and Intel's Pentium chip

As of 1993, the Intel Corporation had sold about $1.8 billion worth of
Pentium computer chips (at the time, Intel's most powerful chip). This chip
served as the brain of approximately 6 million desktop computers. But the
Pentium could not do arithmetic correctly. For instance, the Pentium would
get a wrong answer if you asked it to divide 4,195,835 by 3,145,727.

Some IBM personal computers are built around the Pentium chip, but
IBM in 1993 was planning to market its own microprocessor, called the
Power PC, to compete with the Pentium.

On November 24, 1994, the *New York Times* broke the story that the
Pentium chip made mistakes when it did mathematics. A week later, Intel
published results of an internal risk assessment saying that "average" computer users would get a wrong answer only once per 27,000 years of normal computer use. A "heavy user" might see an error once per 270 years,
according to Intel's risk assessment.

On December 13, 1994, IBM announced it was halting sales of IBM personal computers that depended on the Pentium chip. IBM's risk assessment
had determined that the Pentium chip could cause an error once per 24 days

for average users, and a large company running 500 Pentium-based computers might produce 20 errors per day.

On December 13, the *Times* noted that the two different risk assessments might be colored by financial interests: "But some analysts said yesterday that IBM might have mixed motives in criticizing the Pentium. They noted that IBM was developing its own chip to rival Intel's." (Lewis 1994) In other words, Intel stood to benefit from a risk-assessment conclusion that the Pentium chip was "safe," while IBM stood to benefit from a risk-assessment conclusion that the Pentium chip was "unsafe" (i.e., inaccurate).

If the Pentium chip caused an error once per 27,000 years (Intel's risk-assessment conclusion), maybe Intel wouldn't have to buy back $1.8 billion worth of their Pentium chips. If it caused an error once per 24 days (IBM's risk-assessment conclusion), maybe IBM would have captured a chunk of the desktop computer market for its Power PC chips.

Financial interests are involved in most risk-management decisions, which supposedly are based on objective, scientific risk assessments derived independent of these financial considerations. For instance, decisions on whether highly profitable chlorine production will be legally allowed may depend on assessments of whether incinerators can burn chlorinated wastes with a high enough DRE. Decisions on whether profitable hydroelectric dams should be dismantled if they lie between the ocean and upriver spawning sites of salmon may depend on assessments of whether salmon will continue to exist if the dams remain. A corporation's profits may depend on an assessment of whether arid grassland ecosystems and riparian (stream-side) areas in the public lands of the American west will remain home to a large diversity of native wildlife and plant species when subjected to cattle grazing. The profits of a particular industry may ride on assessments of whether workers' health will be "protected" when the workers are exposed to 400 ppm of ethyl ether.

It's not that every scientist looks to her or his pocketbook and/or employer before drawing up a risk assessment. Even when experts have no financial interests in the outcome, they may assess risks very differently. For instance, in 1990, eleven European governments assembled their most experienced assessment teams to analyze the accident hazards of a small ammonia-production plant (Commission of the European Communities 1991; see Montague 1994f for a brief account of this exercise). The exercise was undertaken to see how closely the eleven different risk assessments of a relatively simple industrial facility would agree. Although no financial outcomes presumably rode on the results, the teams varied in their assessment of the hazards by factors as great as 25,000. Clearly, different risk assessors estimated that the ammonia-production plant posed a much greater or a much lesser risk. An industry or an agency wanting to defend its activities would be more likely to hire a risk assessor known for seeing "less" risk rather than "more" risk in various activities. Many assumptions are present in any risk assessment about the real world. The European risk-assessment report noted that "at any step of a risk analysis, many assumptions are introduced by the analyst, [and] it must be recognized that the numerical results are strongly dependent on these assumptions." This is the flexibility that William Ruckelshaus noted in 1983 for cancer risk assessments. In the case of the Pentium chip, all one side needs to do is find a risk assessor who (perhaps perfectly honestly) estimates that the chip will cause a mistake once per 27,000 years. Who are you to calculate that, instead, the mistake will occur once per 24 days?

In the case of the Pentium chip, Intel and IBM were presenting differing risk assessments. In most situations involving risk assessments, however, only one risk assessment gets prepared. Often this one risk assessment is by a corporation's or an agency's private contractor. If the "expert" or "government" assessment calculates that mistakes or damage will occur once per 27,000 years, how can citizens effectively argue that the damage may occur once per 24 days?

What if a toxic chemical causes prostate cancer in one out of every 24 exposed males rather than one out of every 27,000 males? What if the only risk assessment produced for a decision on whether to permit use of the toxic chemical says that only one out of every 27,000 exposed males will get prostate cancer from the chemical, and so the permit is given?

Often, financial profits and/or political power will be affected by risk-based decisions regarding human activities. Pressure therefore exists to prepare optimistic risk assessments that defend those activities. The health of communities, wildlife, and ecosystems, however, also will be affected by risk-based decisions regarding human activities. Which "side" has a greater opportunity to influence the risk assessments?

3

What Are We Defending with Risk Assessment?

Risk assessment is primarily used to defend unnecessary activities that harm the environment or human health.

It would be hard to find a person who thinks that humans are doing a great job of taking care of the world. Species are becoming extinct at the rate of approximately 50 per day (Wilson 1988), atmospheric ozone is being depleted (WMO 1994), nuclear wastes accumulate (USDOE 1996), some of our rivers stink and others are dried up, urban air is murky, northern lakes are clear but devoid of life because of acid rain, pesticides are in the food we buy and the whales that roam the seas, tropical rain forests are being turned into cattle yards, and we are fighting over the last remnants of ancient temperate forests in the United States. Essentially every organism on Earth is contaminated with numerous human-produced toxic chemicals (Tarkowksi and Yrjanheikki 1989; Muir et al. 1988). We watch dumbly as the human population increases geometrically and the number of molecules of toxic substances it takes to cause havoc turns out to be smaller and smaller.

But almost every human activity that is permitted to damage or kill is accompanied by a formal or informal risk assessment that says "In this case, this activity will have insignificant impacts," or "This amount of this toxic chemical is safe," or "This will not cause unacceptable risk."

Risk assessments vary in their formality. It is exceedingly important, therefore, to detect the essentials of risk assessment even when the words "risk assessment" are not used, or when no complicated quantitative models or formulas are used. If a discussion estimates (however informally) the safety of a hazardous activity or substance without considering the benefits and drawbacks of a decent range of options, it is a risk assessment. (An alternatives assessment, discussing the pros and cons of a range of options, would include an assessment of possible damages or risks of each option. That assessment, however, would be embedded in an assessment of many types of disadvantages of each option, as well as relevant physical, social, economic, and democratic advantages of each option.)

Let us look at some examples of what we are justifying with risk assessment.

Contamination of the Air

"Safety" will be predicted for a hazardous substance or activity via risk assessment even if it is not "safe" in any amount or level. The information in the following example is from a review of studies of the damage done to humans from breathing in small particles (particulate matter) that are released into the air when we incinerate materials (Montague 1994e).

During respiration, we humans can filter out large particles entering our noses and mouths by trapping them on hairs inside our nose, or mucous membranes in our throat, and through other protective features of our airways. Small particles, however, pass by these protective features. They are taken into the deep parts of the lung, where they are deposited on the lungs' surface. Here the toxic materials are given direct entry into the bloodstream.

Combustion (incineration) processes produce many small particles 10 micrometers or less in diameter (a micrometer, abbreviated μm, is a millionth of a meter). Particles that can be inhaled deep into the lungs are often referred to as PM_{10}, meaning particulate matter, or particles, 10 μg in size or smaller. Depending on which materials have been incinerated, these particles are often coated with toxic metals such as mercury or lead or with toxic organic chemicals (e.g., polycyclic aromatic hydrocarbons,[1] which are known to cause cancer). The particles efficiently carry these toxic materials into the bloodstream.

In 1987, the U.S. Environmental Protection Agency (EPA) established a new, stricter standard for PM_{10} in the ambient (outdoor) air: 50 μg/m^3 (micrograms of PM_{10} particles per cubic meter) per day as an annual average, and a ceiling of 150 μg/m^3 to be acceptable only one day a year. That amount of particulate matter pollution, the EPA said, would cause insignificant harm.

But, as it turns out, at least eight studies looking at PM_{10} measurements in the air and death rates of people (see Montague 1994e for a description of these studies) have found increases in death rates whenever PM_{10} particles have been measured in the air, even in amounts below 50 $\mu g/m^3$. One of these, a report in the *New England Journal of Medicine*, described a study of 811 adults in American cities. The study showed that smoking does not explain the increase in human deaths when PM_{10} rises (Dockery et al. 1993).

These studies show that some people will die as a result of breathing air that contains the amounts of PM_{10} that our government permits when homeowners burn wood or industries burn wastes. Joel Schwartz and Douglas Dockery, researchers at the Harvard School of Public Health, estimate that inhaling fine particles contributes to approximately 60,000 deaths in the United States each year. That means particulate-contaminated air is implicated in about 3 percent of the 2 million deaths in the United States each year.

Poisoning of Wildlife

The conclusions of a risk assessment depend on what information is plugged into the risk assessment's calculations. The information in the following example is from a U.S. Fish and Wildlife Service biological opinion about the effects of dioxin on bald eagles along the Columbia River (USFWS 1994).

Bald eagles are currently not reproducing well in the Lower Columbia River, the near-ocean part of the river that flows south out of British Columbia and then flows west between Oregon and Washington to the Pacific Ocean. Most of the bald eagles present along the Lower Columbia River live there year-round. They eat mostly fish, but they also eat some fish-eating and non-fish-eating birds. These fish and birds are polluted with a number of pollutants, among them the extremely toxic dioxin commonly called TCDD (specifically 2,3,7,8-tetrachloro-dibenzo-p-dioxin, abbreviated 2,3,7,8-TCDD). Most of the dioxin that pollutes these fish and birds is stored in their fats. Bald eagles then retain in their fats the dioxin from all the fish and birds they eat. In this way, levels of dioxin contamination in the bald eagles' bodies become even higher than the amounts that were present in the fish and birds they ate.

Certain other species of animals (e.g., mink) have been found to experience difficulty in producing young when they have been contaminated with the amounts of dioxin that have been found in these bald eagles of the Lower Columbia River. This raises concern that dioxin may be part of the reason many of the Lower Columbia River bald eagles are not succeeding in producing enough young to maintain or restore their population.

To make matters worse, certain other chemicals act just like dioxin, except they aren't quite as toxic at the same minuscule levels. These dioxin-like chemicals, including certain dioxins other than 2,3,7,8-TCDD, furans, and polychlorinated biphenyls (PCBs) are handled by an animal as if they were 2,3,7,8-TCDD, and so they cause the same toxic effects, such as developmental damage, reproductive failure, cancer, immune-system damage (O'Brien 1990), and disruption of the endocrine (hormone) system (Hileman 1993).

In 1991, the EPA estimated the amount of dioxin that chlorine-using pulp mills, farms, incinerators, wood-preservative plants, sewage-treatment plants, and other facilities could dump into the Columbia River every day. This "acceptable," permitted amount of pollution, called the Total Maximum Daily Load (TMDL), was based on a risk assessment that predicted that only one person in a million would get cancer from eating dioxin-contaminated fish and drinking dioxin-contaminated water from the Columbia River. (Dioxin also has developmental and reproductive effects at low levels; these were not considered.)

However, humans aren't the only ones eating fish from the Lower Columbia River. Bald eagles in the Lower Columbia River are listed by the Fish and Wildlife Service as a threatened species under the Endangered Species Act. In 1992 the EPA looked at the amount of dioxin pollution it permitted in the TMDL to see if it would kill bald eagles. The EPA did a risk assessment and estimated that the bald eagles would not be harmed.

In 1994 the Fish and Wildlife Service challenged the EPA's conclusion about the "safety" of this amount of dioxins, declaring that, even without consideration of the additional dioxin damage by other dioxins, furans, and PCBs in the Lower Columbia River food chain, the EPA's TMDL for dioxin (2,3,7,8-TCDD) was 10 times too high for bald eagles' safety.

The Fish and Wildlife Service contested the basis of the EPA's judgment, saying that the EPA had looked only at how much dioxin it takes to kill the embryos of adult pheasants injected for only 10 weeks with dioxin. The Fish and Wildlife Service believes pheasants may be more "resistant" to the toxic effects of dioxin than bald eagles, because some animals show toxic effects at lower amounts of dioxin than other animals. The Fish and Wildlife Service said that it would be more realistic to consider how much dioxin it might take to cause the bald eagle, perhaps a much more sensitive bird, to fail to reproduce successfully after eating dioxin-contaminated fish for a lifetime.

The Fish and Wildlife Service also questioned the EPA's assumptions that bald eagles eat only fish, and that the fish in the Columbia River absorb dioxin only from the water. The FWS noted that most of the dioxin the fish accumulate in fact comes from eating dioxin-contaminated organisms in the river.

After considering all these ways that the EPA probably underestimated the amount of dioxin exposure and dioxin effects in bald eagles, the Fish and Wildlife Service concluded that the amount of dioxin the EPA is permitting facilities to dump into the Columbia River is killing some unknown number of bald eagles. The EPA, on the other hand, is claiming that the amount of dioxin they are allowing to be present in the Columbia is not dangerous for bald eagles.

If the Fish and Wildlife Service is correct, then the EPA is indirectly allowing pulp mills and other emitters of dioxin to poison and kill an endangered species, the bald eagle.

Contamination of Groundwater

A risk assessment can be changed, without any new information, so that what might have originally been seen as unsafe will later be declared safe.

In 1985, the drinking water of the eastern Oregon farming area of Ontario was tested for pesticides for the first time. The herbicide dacthal, which is always contaminated with the most toxic dioxin (2,3,7,8-TCDD) was found in seven of seven wells sampled, at concentrations up to 29 micrograms per liter (μg/L, which is equivalent to parts per billion).

At the time, onion growers were applying dacthal on a 10,000-acre area around Ontario at the rate of 6 pounds per acre (Bruck 1986). The soil in this area is well drained, which means that rainwater (and pesticides or other toxic chemicals dissolved in it) easily moves down through spaces between the soil particles. Water is present in the ground at only 9 to 11 feet below the surface of the soil in the Ontario area. The drinking water is quickly contaminated in this shallow aquifer.

In the next year, 1986, sampling found 54 of 81 wells contaminated with dacthal, at concentrations up to 431 µg/L (Pettit 1987).

Was well water containing this amount of dacthal "safe" to drink? In 1986, the only existing "standard" for dacthal was a 1982 EPA draft "health effects guidance" that set a threshold of concern at 500 µg/L (USEPA 1982). Such a standard implies that if you are drinking water with less than 500 µg/L of dacthal you are "safe" (or your chances of getting cancer won't be greatly increased) but that drinking water with more than 500 µg/L dacthal might cause health damage (or an unacceptably high chance of cancer).

This 500 µg/L standard was based on a two-year rat study, conducted in 1963, that had found effects at all doses and "a variety of tumors" in the exposed rats (USEPA 1982).

The standard had been calculated as if the lowest dose had not caused any adverse effects (which was not true), the tumors had not been observed (which was not true), and a 20-kilogram (44-pound) child were drinking the water. And the standard was calculated using a "safety factor" of a 1000-fold reduction from the lowest dose in the rat study (USEPA 1982). This safety factor was composed of three safety factors of 10 (i.e., $10 \times 10 \times 10 = 1000$). One safety factor of 10 often used by risk assessors assumes that humans are 10 times more sensitive than an animal that has been tested (in this case, a rat). In fact, humans may be more than 10 times as sensitive as a rat. A second commonly used factor of 10 assumes that some humans may, because of genetic variability, be 10 times more sensitive than most humans. In fact, some humans may be more than 10 times as sensitive than most humans. The third factor of 10 the assessors used is commonly used when the animal experiment upon which a standard is based is inadequate.

Compared to standard experimental designs, the dacthal study certainly qualified as inadequate. An experiment like this usually includes several groups of healthy rats, one of which is untreated, and other groups that are exposed to differing amounts of the toxic chemical. In this experiment, however, a lung infection was present in all groups of rats in the experiment, so the rats were dosed with antibiotics throughout the experiment. Presence of a lung infection and/or exposure to antibiotics could have affected what happened to the rats exposed to dacthal.

Moreover, both experimental rats (those fed certain amounts of dacthal) and control rats (those not fed any dacthal) died at a high rate during the experiment. Both the high mortality rate during the study and the sacrifice of some other rats partway through the study reduced the numbers of animals from which the experimenters could draw conclusions about the effects of dacthal (USEPA 1982).

One year after dacthal had been found at 431 µg/L in the groundwater of eastern Oregon (approaching the 500 µg/L standard based on the inadequate study), the EPA issued a new "health advisory" for dacthal in drinking water that set a standard of 3500 µg/L (USEPA 1987a).

Why did dacthal suddenly look 7 times "safer" now? Had the EPA received new, more adequate studies since determination of the 1982 standard? No. The EPA was using the same inadequate 1963 rat study to calculate the standard. The EPA merely plugged different numbers into the formula. The new formula assumed that only 70-kg (154-lb) adults, not children, would drink the dacthal-contaminated water, and it applied a 100-fold safety factor rather than the 1000-fold safety factor. The EPA had dropped the factor of 10 that provided caution because the study had been inadequate. These two factors, offset slightly by other minor changes (e.g., the adult would drink 2 liters of water a day rather than the child's 1 liter) resulted in a standard that made dacthal look 7 times "safer" than it had in 1982.

The 1987 dacthal standard still treated the risk of cancer as zero, even though tumors were found in the treated animals. Neither standard mentioned that dacthal is contaminated with hexachlorobenzene (Wapensky 1969), which causes cancer (Cabral et al. 1977) and which has been found in the blood of certain human populations at levels near those that cause liver and kidney damage in fish (*Pesticide Chemical News* 1976).

Additionally, the 1987 health advisory did not mention that dacthal had been found in 1985 to be contaminated with 0.16 part per billion (ppb) 2,3,7,8-TCDD, the most toxic known dioxin (USEPA 1985a).

This incident shows how easily risk assessments can be (and often are) altered to produce results desired by risk assessors or risk managers. With so many numbers in a risk assessment being "guesses" or "assumptions," it becomes difficult or impossible to challenge the alteration of individually crucial factors in a risk assessment.

"Acceptable" Nuclear Radiation

A certain amount of a hazardous substance can be deemed "safe" by regulators at one time but later be understood to be harmful or even lethal.

The "standard" for how much exposure to nuclear radiation will be "protective" of nuclear workers has shrunk from 156 rem in 1925 to the current U.S. standard of 5 rem per year, adopted in 1968 (see figure 3.1).[2] In 1990, the National Research Council's fifth "BEIR" (Biological Effects of Ionizing Radiation) report (NRC 1990) asserted that radiation is almost 9 times as damaging as was estimated in the first BEIR report, issued in 1972.

In 1990, on the basis of the scientific evidence that the 1968 worker standard of 5 rem per year was not protective, the International

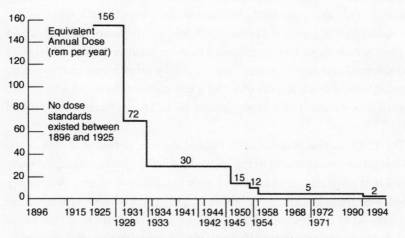

There is no single set of radiation protection standards. This graph is based on recommendations, sometimes different, published by U.S. and international groups concerned with radiation protection. They have been translated into a single, consistent set of numbers and measurement units for the purpose of this summary.

Figure 3.1
The evolution of health-protection standards for nuclear workers. Source: USDOE 1995.

Commission on Radiological Protection[3] recommended a worker standard of only 1 or 2 rem per year. As of 1995, however, the U.S. standard remains at 5 rem per year.

John Gofman, a former Director of the Biomedical Research Division of the Lawrence Livermore National Laboratory, contends that even the fifth BEIR report underestimates the damage caused by low-dose radiation. He offers evidence that the National Research Council wrongly relies on non-human evidence to arrive at its estimate of the risk of low-dose rates of radiation. He points to human evidence that there is no amount of radiation exposure that does not cause adverse effects in at least some individuals (Gofman 1991). Moreover, Gofman contends, the human evidence indicates that the risk of cancer may be higher per dose for low doses of radiation than for moderate or high doses. In other words, a given unit of radiation coming in low doses may be more potent at causing cancer than the same unit of radiation coming in a moderate or a high dose.

The issue is not academic. The BEIR V data would predict 1.4 excess deaths of cancer if 1000 workers were exposed to 5 rem per year (i.e., 1400 deaths per million exposed workers). Gofman's data would predict 5.3 excess deaths per 1000 workers (i.e., 5300 deaths per million exposed workers). Gofman would predict almost four times as many cancer deaths when workers are exposed at the current worker "standard" for radiation than would the BEIR V group.

On June 1, 1995, Genevieve Roessler, a radiation biologist and a long-time consultant to nuclear utilities, co-directed a risk-assessment workshop for the Hanford Advisory Board, a citizen advisory board that provides broad-based recommendations to the Department of Energy regarding the "cleanup" of the massively contaminated Hanford Nuclear Reservation in south-central Washington, adjacent to the Columbia River. At this workshop, Roessler referred to the National Research Council's BEIR V risk assessment of low-dose radiation as "conservative." By this she meant the risk assessment probably overestimates the risks of radiation. She never mentioned the evidence marshaled by John Gofman that the new BEIR V risk assessment may in fact significantly underestimate the risks associated with low-dose nuclear radiation.

It is rather remarkable how commonly government agencies and industry pronounce their risk assessments "conservative," as if they knew the true risk; as if the many uncertainties in any risk assessment don't reveal that the risk assessment, in fact, could be significantly non-conservative; and despite the fact that historical records reveal countless instances in which substances or activities were found to be more dangerous than originally estimated, when more scientific information was gathered.[4]

Extinction of Species

Risk assessment is used to predict not only death or damage to humans or wildlife from human activities but also the extinction of species. In March 1993, the U.S. Forest Service prepared an assessment of the environmental risks of cutting more old-growth forests in the western half of Washington, Oregon, and northern California (USFS 1993b). These forests contain large Douglas fir, cedar, and other trees that are hundreds of years old, as well as fallen logs, standing dead trees, and certain plants and animals that at least partially depend on the old-growth community. One of the rare species that depend on old growth for habitat is the spotted owl (*Strix occidentalis*). More than 90 percent of such ancient native forests have been destroyed already in this section of the Pacific Northwest.

The authors of the 1993 risk assessment said they couldn't assess the probability of survival of the spotted owl by a formula, because neither the formula (model) being developed by the Forest Service nor maps of spotted owl habitat were ready. Later, when these Forest Service maps and a model for calculating spotted owl survival potential were completed, the same authors prepared a second report (USFS 1994b), which said that it would be years before they knew enough about the spotted owl to prepare estimations of survival using the model.

Meanwhile, in July 1993, a team of Forest Service and Bureau of Land Management personnel devised eight different alternatives (or "options"), each successive option proposing more logging of old-growth forests. President Bill Clinton, however, requested that a ninth option proposing even more logging of old-growth forests be developed.

In February 1994, the Final Supplemental Environmental Impact Statement (USFS 1994a), regarding the environmental consequences that might occur if Option 9 were implemented, was prepared. The authors of this statement had prepared a risk assessment estimating the probability of survival of various animal species that depend in some way on old growth.

The EIS noted that a panel of scientists had concluded that Option 9 would result in an 83 percent likelihood of habitat sufficient to support a viable spotted owl population for the next 100 years.

What this EIS didn't say was that this prediction was based on pooling the estimates of four panel members, two of whom gave only a 70 percent chance of the spotted owl's survival. One had predicted an 83 percent chance, and a fourth, who has done professional work for the timber industry, had given a 100 percent chance of survival. Under oath and under questioning by Sierra Club Legal Defense Fund lawyers, both of the panelists who guessed the higher probability of spotted owl survival admitted that they did not have the expertise to predict species extinction or population viability (i.e., the capability of populations of a rare animal to survive, given particular adverse conditions in their habitat) (Sierra Club Legal Defense Fund 1993).

Finally, this panel and other panels that were to estimate chances of extinction of the spotted owl and other old-growth-dependent species for the EIS were told they could not consider the impacts of activities that would likely be happening on private lands adjacent to the national forests of Option 9, even though such activities (e.g., logging, building of roads, constructing buildings on the land) might lower the chances of survival of at least some of the species.

In fact, the panel looking at survival of old-growth-dependent snails (mollusks) refused to ignore the cumulative effects of human activities on state-owned and privately owned land adjacent to the National Forests. They concluded that only Option 1, the option that involved the least logging of old-growth trees, would avoid destruction of any populations of the rare mollusks.

The hodgepodge process that was used to estimate risks of extinction did not, however, stop President Clinton from adopting Option 9.

This example shows that decision makers will use scientifically indefensible and economically and politically influenced risk assessments to justify desired activities even when that may mean extinction of numerous species.

Addition of Toxic Chemicals to Our Food

When scientists and risk assessors become convinced that risk assessments are not considering important information (e.g., infants' special physiology, cultures with particular diets, habits of particular species) they will strive to make the risk assessments more complicated and more data-rich (i.e., full of more information). Meanwhile, these scientists and risk assessors generally fail to ask whether the hazardous activity that is being assessed is even necessary or ethical.

In 1988, Congress asked the National Academy of Sciences to establish a committee within the National Research Council to study scientific and policy issues surrounding pesticide residues that the federal government permits to be present on foods eaten by infants and children. Five years later, the committee published its findings in a book called *Pesticides in the Diets of Infants and Children* (NRC 1993). The book noted that the levels of pesticides currently "permissible" for adults are not necessarily "safe" for children for a number of reasons.

Children or infants may absorb a greater proportion of a pesticide inside their body, and they may not break down (detoxify) or get rid of (excrete) pesticides as readily as adults. Even more critical, the committee wrote, infants and young children may eat much more of certain foods (e.g., apples via apple juice and applesauce) per body weight than adults, and so may receive more of certain pesticides than was "expected" for an average adult.

The committee's recommendations are numerous, and they focus on producing more comprehensive risk assessments so we won't allow "too much" pesticide residues on foods that infants and children eat.

For instance, the committee recommended that studies of the effects of pesticides on laboratory animals should focus on health damage effects to which infants and children are particularly vulnerable during certain periods, such as damage to the immune-system, to the developing reproductive system, and to the nervous system. When such information is missing (as it is most of the time), extra "safety factors" should be plugged into the numerical risk-assessment formulas. These safety factors would result in a reduction of the amount of pesticide that would be considered "safe." The committee recommended that we learn more about what foods infants and young children eat, and in what amounts. They recommended more frequent testing of the amounts of pesticides that are present on foods in the store. They recommended that we learn more about which chemicals transform into other, possibly more toxic chemicals as food is processed (for instance, when tomatoes are heated for canning). They recommended adding the amounts of different pesticide residues on a food together if they cause the same adverse effects (for instance, a particular interference with the nervous system). Adding similar pesticides together would result in lower allowable amounts for any one of these pesticides. They also recommended that we learn more about how cancer rates are affected by exposure to a cancer-causing pesticide at a certain time in life.

Overall, the committee's recommendations "support the need to improve methods for estimating exposure and for setting tolerances [acceptable amounts of a pesticide on any given item of food] to safeguard the health of infants and children" (NRC 1993).

The committee did not address the issue of toxic "inert" ingredients— ingredients in a pesticide formulation that don't directly kill or repel the pest. Approximately 1450 "inert" pesticide ingredients are registered for inclusion in pesticide formulations to dissolve the pesticide (solvents), spread it evenly out over leaves or insects and keep it in place (spreaders-stickers), keep it in suspension in the tank (emulsifiers), help the pesticide coat the surface of a leaf or insect or other "desirable" place, and so on. The toxic effects of about 1000 of these 1450 ingredients have not been studied; they are called "inerts of unknown toxicity" by the EPA (USEPA 1984b). The committee did not address the concern that a risk assessment may essentially be meaningless if only one ingredient in a pesticide formulation has been tested but the untested ingredients are in fact highly toxic.

More important, the committee did not recommend producing food organically, or altering agricultural methods to reduce dependency on pesticide use. It did not recommend allowing zero amounts of a particular pesticide on children's food if that pesticide is not one that "needs" to be used to produce the crop. It did not question whether pesticides "have" to be on children's food, nor did it call for study of the availability of alternatives to using pesticides. It simply made recommendations for figuring out how much pesticide we should permit on children's food.

Interestingly, the Maine Board of Pesticides Control (1997) took a different stance with regard to Maine registration of forage corn-seed products with an insecticide incorporated within their genetic structure. These corn seeds have been genetically modified by the Dekalb Genetics Corporation so that they produce a bacteria, *Bacillus thuringiensis*, that will kill the European corn borer if it feeds on stored corn seeds. However, organic farmers argue that its widespread use will only encourage the development of insect resistance to *Bacillus thuringiensis*, eventually reducing its effectiveness when needed. The Board, citing a lack of need, refused to register this genetically altered corn.

In the end, will we best protect our children by very carefully calculating how much pesticide to feed our children, or will we best protect our children by feeding them foods grown without pesticides? This is not the kind of question the National Research Council's committee of experts asked. Most risk assessors don't ask about the necessity of undertaking the hazardous activities they assess.

Reduction of Wildlife Habitat

A risk-assessment approach to an issue can be so informal that the phrase "risk assessment" is never even used. Such a risk assessments may have no complicated tables, numbers, or formulas, and thus may not "look" like a risk assessment.

The information in this section is from a Forest Service environmental assessment of a proposal to build tourist roads and developments on a canyon rim (USFS 1993a) and from a review of elk-management research (Christensen et al. 1993).

In 1994 the managers of the Wallowa-Whitman National Forest in northeast Oregon wanted to build a series of "motorized tourist developments" on the southern part of the western rim of Hells Canyon National Recreation Area. At the present time, this part of the western rim is rarely traveled by vehicles, except during the fall hunting season. The Forest Service proposed to pave 6 miles of road for tourist buses, RVs, and other motorized vehicles; to widen and "improve" additional miles of gravel road; to construct four overlooks with parking lots, rest rooms, and interpretive displays; and to construct two trailhead campgrounds with parking lots and livestock-loading and tie-up facilities. These developments would have been constructed immediately adjacent to the Hells Canyon

Wilderness and would have provided another setting for people in motorized vehicles to look out over Hells Canyon, the deepest river-carved canyon in North America.

The trouble is that the rim area that would have received the greatest development currently is crossed by migrating Rocky Mountain elk, mule deer, and bighorn sheep as they move down into Hells Canyon to the east for the winter and back up into the Wallowa Mountains to the west in the summer. An adjacent ridge that also would have been "developed" is a spring and summer calving meadow for elk. The Congressional law that created the Hells Canyon National Recreation Area (Public Law 94-199, 94th Congress, December 31, 1975) requires the Wallowa-Whitman National Forest to provide "protection and maintenance of fish and wildlife habitat."

In order to justify these tourist developments and meet the requirements of the Hells Canyon National Recreation Area Act, Forest Service personnel described two "needs" for the project: to meet the recreational need for improved recreation access and facilities and to meet the need for increased elk security.

To meet recreational "needs," the Forest Service proposed to build the complex of tourist developments. To increase elk security, the Forest proposed to "close" 71 miles of dirt logging roads that are on the few small areas of flat land adjacent to the rim-road developments. These roads are rarely used except during hunting season, when hunters camp nearby. The Forest Service indicated that these road closures would reduce the "open road density" in the area from 4.14 miles of road per square mile to 1.69 miles of road per square mile. They claimed that the combination of tourist developments and road closures would, overall, "improve elk habitat."

This, then, was their "informal" risk assessment: Building 10 miles of paved and gravel tourist facilities and roads, while closing 71 miles of backcountry dirt roads, will result in reduced road density and, therefore, "improved elk habitat." (One person who read this plan noted that this is "like closing all the roads in town and routing all traffic onto a superhighway through the playground to protect the kids and adding a few casinos and theme parks to bring in more traffic.")

The reference to road density relates to a growing body of scientific information showing that grizzly bears, wolves, elk, wolverine, and some other

wildlife have difficulty surviving in areas where humans have much motorized access. One way to assess the risks of roads is to calculate the numbers
of miles of open road per square mile of wildlife area. Using this risk-assessment method, the Forest Service claimed that by reducing road density it
was improving the elk habitat.

However, the analysis of wildlife survival and road density is becoming
increasingly sophisticated, and factors other than road miles per square mile
are being recognized as important for elk habitat. These factors include
number of hunters per unit area; size, shape, and distance from roads of
"security areas" of forest cover; location of the roads (away from or on a
migratory route); and terrain (rough or flat).

For instance, research shows that if there are 1.69 miles of road per
square mile in an area and one hunter comes to the area, an elk has a 30 percent chance of being killed that season. If 37 hunters come to the same area
with the same road density during the season, the elk has approximately a
70 percent chance of being killed (Christensen et al. 1993). In other words,
the number of road users is to be considered, as well as the sheer density of
roads.

In its informal "risk assessment" of the proposed tourist developments
and road closures, the Forest Service failed to consider any factors other
than miles of road. For instance, it failed to consider whether more hunters
would come to the area if the road were paved, or how many tourists would
be coming to the facilities.

Even though this Forest Service risk assessment was informal (i.e., "Elk
habitat will be improved if we close some roads and build some others")
and wasn't labeled a "risk assessment," it suffers from the same arbitrari-
ness of assumptions and lack of alternatives (e.g., an alternative that emphasizes non-motorized use of the rim by visitors) that mark most formal risk
assessments.

Denying the Right of Nations to Establish High Health-Protection Standards

This example shows how risk assessments are being used at the international level, and how one risk assessment can affect many countries. The
sources of the information are Kelly 1994 and O'Brien 1992.

In the late 1980s, industry began to see how trade agreements among countries (called multi-lateral trade agreements) could be used to influence toxic substances regulatory standards within countries. The World Trade Organization uses the word "harmonization," as did the General Agreement on Tariffs and Trade before it. The idea of harmonization is that countries trading with each other (for instance, Mexico, the United States, and Canada under the North American Free Trade Agreement, and 123 governments under GATT) should all agree to abide by the same legal limits for imported toxic materials. Harmonization would apply to toxic substances like pesticide residues on imported food, paint, and asbestos. Harmonization would assure a manufacturer in one country that it could export, sell, or import toxic products to or from anywhere without having to face particularly stringent standards in certain countries.

The various agreements don't say outright that no country could have more protective standards than another country. However, the agreements call for adoption of international "standards" for pesticide residues on food that is exported or imported, and for harmonization of import rules. The agreements say that if a country wants to establish more protective toxic substances rules than other countries it must defend these rules with "risk assessments." Any other protective rules would be considered a "trade barrier" to companies in other countries, and a country would be in violation of the trade agreement.

For instance, under the North American Free Trade Agreement (NAFTA), if the United States allowed less DDT on imported apples than the international standard, it would have to demonstrate that it had considered both the use of international methods of risk assessment and how to minimize negative trade effects. In other words, the United States would have to show that it had considered ways to minimize financial discomfort to agricultural corporations who wanted to export apples with high amounts of DDT on them to the United States.

Under GATT, a country that wants to develop a more stringent or protective rule is required to consider the "exceptional character of human health risks to which people voluntarily expose themselves" (GATT 1993). Presumably this means that a country developing some limit for a cancer-causing pesticide would have to take into consideration the cancer risk some of its addicted citizens "voluntarily" assumed by smoking three packs of cigarettes a day. Then, if the pesticide causes less cancer than smoking three packs of cigarettes a day, the country would have a harder time defending its decision to limit that cancer-causing pesticide.

The issue of who develops international toxic substances standards is also troubling. International standards for pesticide residues acceptable on foods, for instance, are developed by the Codex Alimentarius Commission, a subsidiary body of the United Nations Food and Agricultural Organization. Codex Alimentarius is strongly influenced by private chemical-industry and food-industry groups. Codex Alimentarius does not have any process for public input or scrutiny of its decision-making processes.

This is clearly a bonanza for industry groups: They get to establish lax pesticide residue limits on foods internationally through the secretive processes of Codex Alimentarius, then they impose these lax limits on other countries through multi-lateral trade agreements. An industry in one country can then get its country to challenge more protective standards in another country. The country that tried to develop more protective standards may then "lose" the right to be more protective through secretive processes for resolving international disputes that are also secretive. Democratic or other life-protecting processes within that country will therefore have lost out to world government by industry-driven, secretive risk assessment.

Box 3.1
How Agencies and Companies Can Manipulate a Risk Assessment So That It
Will Downplay the Dangers of a Dangerous Activity

Downplay estimates of hazard

Consider only one chemical or activity at a time, ignoring cumulative or synergistic effects of multiple chemicals or activities.[1]

Consider only hazards that have been repeatedly "proved" in controlled experiments with healthy animals of one genetic strain.

Consider as few hazards as possible, perhaps only cancer, or extreme damage.

Consider evidence of hazard from studies of small numbers of people or animals exposed, with a short time period since the exposure, and only moderate exposure.

Discount harmful effects experienced or observed and reported by local individuals and communities as "anecdotal."

Downplay estimates of exposure[2]

Estimate average exposure of healthy, suburban, white adults, ignoring children, ill people, old people, ethnic groups that have eating habits or undertake other cultural practices that are different than many Americans, or people who are having to take immune-suppressing drugs.

Avoid considering instances where exposure has been found to be unusually high, or where spills or other kinds of unusual occurrences have resulted in more intense exposure.

Assume a relatively short time of exposure.

While writing the exposure scenario, do not work with community people or other people who are familiar with the real world (e.g., their neighborhood, or the local ecosystem being considered) and who have observed high exposure or damage there.

Use complicated mathematical models or formulas that can only be analyzed in a complicated computer program that community groups cannot easily gain access to or understand.

In order to estimate expected exposure, use proprietary (non-public) mathematical models or formulas that have been used by private consulting companies when they estimated insignificant risk for similar activities or substances.

Present estimates and unfounded assumptions as factual certainties.

Downplay risk

Ignore such risks as damage to community life, the practice of democracy, the spirit of people, future generations, or traditional Native American practices, because a number cannot be assigned to them.

Ignore risks of "unimportant" impacts such as immune suppression, headaches, anxiety, or wildlife not being able to migrate on accustomed routes.

Explain the risks in highly technical terms.

Compare the risk to other, "voluntary," risks (e.g., smoking), or natural risks (e.g., hurricanes). (Explain, for instance, that more people die from hurricanes than from DDT.)

State that we are all living longer lives now, so there is no problem here. (This ignores that fact that more people are getting cancer (Miller et al. 1992) and the quality of life is lowering for many.)

Do not discuss whether the risks are necessary or whether they could be avoided entirely through reasonable alternative behaviors.[3]

source: Mary O'Brien

---

1. This is perhaps the second most important step.
2. Exposures are almost always unknown in reality.
3. This is the most important step.

4

When Scientists Shut Their Eyes: Pretending That the Safety of Hazardous Activities Can Be Estimated

Most risk assessments are unscientific because they cannot estimate how much of a damaging activity poses no risks or "insignificant" risks.

Imagine someone assessing how many Bloody Marys it is "safe" to drink in one day, but measuring only the tomato juice and not the vodka. The assessor might conclude that drinking a gallon of Bloody Marys (i.e., tomato juice) would pose no danger at all. If someone enthusiastically relied on that risk assessment, the effects could be fatal.

Industry and government enthusiastically rely on the conclusions of many such partial risk assessments. The results are generally destructive, and often fatal.

Scientific studies can be very good at determining some of the adverse effects of a particular activity (e.g., clearcut logging) on an ecosystem or some of the effects of a particular substance (e.g., dioxin) on an organism. But scientific studies are not good at determining how much of a stressful activity can be "handled" by a particular ecosystem or how much dioxin is "safe" for some animal out in the world.

The major scientific flaw of most risk assessments is that their goal is not really to warn of damages that might occur, as the word "risk" would imply. Instead, they are developed primarily to estimate how much of a hazardous substance or activity does not pose risk for living organisms out in the real world. (See box 4.1.)

Scientific studies and scientists can help us understand some of the risks and damages caused by particular substances and activities, but these studies and scientists are generally not able to tell us how much of a dangerous substance is safe for us or how much of a harsh activity can be handled by an ecosystem.

Box 4.1
"Safe" and Some of Its Variants

Safe[1]	Clean bill of health
Insignificant[2] risk	No inherent danger
Acceptable daily intake	No threat to public health and safety
Reference Dose	No one was harmed
De minimis risk	Protective of public health
Acceptable risk	Less dangerous than aspirin
Below the threshold for effects	Ample margin of safety

source: Mary O'Brien

1. "Safe," according to *Webster's Third International Dictionary*, means "1. Free from harm, injury, or risk. 2. Secure from threat of danger, harm, or loss. 3. Affording protection from danger. 4. Not threatening danger."
2. "Insignificant," according to *Webster's*, means "having no importance."

This chapter first explains how scientists demonstrate some of the damages that might be caused by certain substances and activities. It then explains why scientists and risk assessors cannot generally use information on risks to estimate how much of a hazardous substance or activity is "safe" or will cause only "insignificant" damage.

Scientists Can Demonstrate Risks

Example: How Scientists Discovered That Dioxin May Cause Endometriosis in Monkeys

In 1977, an experiment was undertaken to study adverse effects in offspring of monkey mothers who had been exposed to dioxin (Rier et al. 1993). For the purposes of this study, three groups of female rhesus monkeys were treated differently. One group of six monkeys was kept free of exposure to dioxin, while the two other groups of seven monkeys each were fed different amounts of dioxin, each group for five years (1977–1982): One group was fed 5 parts dioxin per trillion parts monkey body weight (5 ppt); the other was fed 25 ppt.

In 1989, 7 years after the feeding experiment ended, one of the female monkeys who had been treated with 25 ppt of dioxin died, and an autopsy showed that she had been suffering from endometriosis, a painful disease in which pieces of uterine tissue grow outside the uterus. In 1991 a second monkey in the 25-ppt group died, and in 1992 a third monkey in the 25-ppt group died; both were found to have severe endometriosis. Upon examination of the 17 other female monkeys still alive, and counting the three who had died with endometriosis, the following was noted: Three of seven monkeys exposed to 5 ppt dioxin (43 percent) had moderate to severe endometriosis, and five of seven monkeys exposed to 25 ppt dioxin (71 percent) had moderate to severe endometriosis. Two of the six monkeys with no exposure to dioxin (33 percent) had endometriosis, neither of them severely. (Of 304 monkeys housed in the primate center, 30 percent were diagnosed with some degree of endometriosis; this would seem to indicate that about 30 percent of monkeys, at least caged monkeys, develop at least mild endometriosis.)

This is powerful information, because the larger the dose of dioxin the more frequent is the occurrence of endometriosis. This would suggest that dioxin plays a role in causing endometriosis to occur.

Although this experiment was undertaken with a non-human species, the experimental evidence leads to concern that contamination of humans with dioxin and ongoing exposure to dioxin may be playing a role in causing some of the endometriosis that is being experienced by women. Currently an estimated 5 million women in the United States suffer from endometriosis (Endometriosis Association 1993).

One of the risks of exposure to dioxin, then, may be an increased chance of suffering endometriosis.

Example: How Scientists Are Showing That Cattle Grazing within Western Forests Can Lead to Sick Forests

Sometimes scientists can study effects of a substance or activity out in the real world rather than in the laboratory. This has happened in studies of what occurs when cattle are excluded from a western U.S. grassland area that is similar to a grassland area where cattle grazing continues. An example of this is a study in which two researchers looked at three similar plateaus in Utah: Horse Pasture Plateau, Church Mesa, and Greatheart Mesa (Madany and West 1983). All three have ponderosa pine forests. Horse Pasture Plateau was grazed by livestock from the late 1880s until

1960, whereas Church and Greatheart Mesas were never grazed by livestock because of steep cliffs. None of the three pastures burned between
1892 and 1964, so presence or absence of livestock grazing was the only significant environmental difference between Horse Pasture Plateau (grazed
by livestock) and the two mesas (not grazed by livestock). The researcher
found that during the period when Horse Pasture Plateau was being grazed
(1880–1960) the appearance of young trees increased by a factor of 10,
while during the same period the appearance of young trees remained the
same on the two mesas. Eventually, for every mature ponderosa pine on
Horse Pasture Plateau there were 598 young trees, while on the two mesas
the proportion remained 0.8 young tree per mature tree.

This is powerful information, because (1) the only apparent difference
between the two areas is whether cattle have been grazing the area and (2)
the time period for observations is quite long (80 years), which reduces the
chance that the observations could be due to short-term events such as
unusual weather or an insect outbreak.

The information is significant because one of the reasons being given at
the present time to log western public forests (i.e., national forests) is that
too many trees are growing on many sites, causing weakened growth of
individual trees, pest and disease outbreaks, and fire hazards. Many scientists have been blaming this "overstocking" of trees exclusively on the
"Smokey the Bear" policy of putting out all fires, including lightning-caused
fires. Before that policy, natural fires would have destroyed many younger
trees in a forest while allowing the more mature trees to survive. The fires
helped keep forests open and healthy.

What this three-plateau study shows, however, is that current livestock
practices may also be causing the same overstocking problem, and that if
we continue grazing cattle under forest trees we will continue producing
unhealthy, overstocked forests, even if we also allow occasional fires. When
cattle graze the ground between the trees of an open forest, they eat grasses and forbs (broad-leaved herbaceous plants) that would have competed
with tree seedlings (Belsky and Blumenthal 1995).

Therefore, it is not only fire that keeps forests open and uncrowded by
"weeding" out weak young trees; grasses and forbs also control the number of tree seedlings by choking them out. Without fire, or with cattle
removing the tree-limiting grasses and forbs, a crowded and disease-prone

forest builds up. Then, when fire does come, it can be much more destructive than it would have been in a more open forest with fewer trees.

One risk associated with grazing within western forests, then, is that the forests will become unhealthy and overcrowded with trees. However, that is not a risk that is usually included in environmental assessments accompanying government permits for private ranchers to graze livestock in national forests and other forested public lands.

Scientists Generally Cannot Show Safety

As was noted above, both laboratory and real-world experiments can demonstrate potential, likely, or even inevitable damage (which are all lumped together as "risks") of particular activities or substances. However, the purpose of most risk assessments is to estimate levels of a hazardous activity (e.g., emitting dioxin or livestock grazing) that will cause no damage or no significant adverse effects out in the world. The scientists' skills and training for showing effects in controlled experiments are simply inadequate for showing that those activities are safe in a complex world.

Let us return to the Bloody Mary analogy, but let's change the story a bit. This time a laboratory scientist is studying adverse effects, if any, that occur to people who drink a Bloody Mary (this time, both tomato juice and vodka). The people in the experiment have eaten dinner, but they have not drunk alcohol for a week. The researcher may conclude that there are few adverse effects, none of them "significant," from drinking a Bloody Mary.

However, in the real world, outside the experiment, that same Bloody Mary may render a person unconscious if he or she has already been drinking all evening and hasn't eaten much food.

Making estimates only on the basis of the experiment, the researcher could say "Drinking a Bloody Mary is safe." A Bloody Mary is not safe, however, for someone who recently has been consuming too much alcohol.

Like a person whose body has been exposed to too much alcohol, a forest may be experiencing a multitude of stresses as a consequence of human activities (box 4.2). While a single activity, such as grazing a few livestock in a forest, may not be terribly detrimental to a forest that periodically

Box 4.2
Some Impacts That May Be Experienced Cumulatively by a Forest at Any Given Moment

> Each of the following places stresses on at least some plant or animal species, upon which other plant or animal species may depend:
>
> Increased UV-B light (from ozone layer thinning)
> Acid rain
> Heavy metal deposition
> Global warming
> Mining
> Pesticides
> Other chlorinated toxic chemicals
> Other endocrine-disrupting chemicals
> Exotic insects, plants, and animals
> Livestock grazing
> Roads
> Humans in motorized vehicles (e.g., snowmobiles, jet boats, all-terrain vehicles, automobiles)
> Non-motorized recreation (e.g., hunting, fishing, hiking, swimming, rock climbing)
> Soil erosion
> Fire suppression
> Clearcutting
> Reduced habitat acreage
> Isolation from other populations by surrounding "developments" (e.g., towns)
> Airplane overflights; other noise
>
> *source: Mary O'Brien*

burns, the combination of suppressing fires and grazing numerous cattle and spraying insects may be too much for the forest. In such a situation, the grazing of cattle cannot be said to be "safe" for the forest. The results of an experiment in which a healthy forest is grazed lightly by cattle, then, may not be relevant to a forest in which many other forest-stressing human activities are also taking place.

This quandary of estimating what amount of something is "safe" occurred to me when I was in Santiago, Chile, as a member of a United Nations committee looking at alternatives to using methyl bromide. Methyl bromide is a gaseous pesticide that, after use, escapes into the stratosphere,[1] where it contributes to the destruction of the ozone layer, much like chlorofluorocarbons (CFCs). Countries cooperating to phase out the use of ozone-depleting chemicals such as CFCs and methyl bromide had formed this international committee.

Because of the nature of the stratosphere and the temperature differences between the hemispheres, the ozone layer has been thinned most in the southern hemisphere, although even northern countries are now experiencing thinning too. Many living organisms on Earth need a thick ozone layer to block certain of the sun's ultraviolet rays (called UV-B light) from reaching them. These UV-B rays harm the immune systems of many organisms, cause deadly skin cancer, and damage eyesight, among other effects (UNEP 1991). Since Chile is in the southern hemisphere, it is being exposed to these effects more than northern countries such as the United States (Nash 1991).

Santiago, where the meeting was being held, is a city of 5.5 million people in middle Chile, surrounded by spectacular mountains of the Andes and the Cordillera de la Costa. Santiago is subject to numerous simultaneous stresses on living organisms. For instance, during this particular meeting of the UN committee, several neighborhoods were being aerially sprayed with the organophosphate pesticide malathion to combat a fruit fly infestation ("Santiago sur intoxicado con el bombardeo de insecticidas," *El Siglo Diario*, March 4, 1994). One woman died immediately after the spraying, perhaps in reaction to exposure to the pesticide, and several people temporarily lost some feeling in their hands and arms. A number of neighborhood groups had tried to prevent the spraying.

One afternoon during my visit, I walked in Santiago for several hours with a Spanish woman who works to help provide services for the thousands of Santiago's children who live on the streets, without families, or enough food, or adequate shelter.

During the week of my visit, in March, the mountains that surround the city were never visible from the city, because of air pollution. Although March (i.e., autumn) is considered a better time of year than winter for air quality, my eyes stung and my head ached whenever I was outdoors. I asked myself how a risk assessor could estimate the "safe level" of exposure to malathion—an immune suppressant (Rodgers and Ellefsen 1992) and a neurotoxin[2] (Cremlyn 1991)—in the aerial-spraying program. What if an exposed Santiago citizen's immune system has been compromised prior to the spraying by (1) increased UV-B radiation (Longstreth et al. 1991) because of the ozone thinning; and this citizen is a street child with (2) poor nutrition (i.e., potentially with reduced immune-system resiliency), who is (3) breathing air particulates to which cling toxic chemicals (e.g., immune suppressants and neurotoxins) from industry and traffic?

All risk assessors are in the same position as a risk assessor trying to estimate how much malathion exposure is "safe" for a poorly fed street child in the southern "ozone hole" city of Santiago. Risk assessors can't account for all the conditions in the real world. They can't account for all the toxic substances already contaminating and affecting people and other species out in the world. They can't account for individual sensitivities. They therefore simply put on blinders and proceed full steam ahead, estimating the safety of some substance or activity as if the real world doesn't exist.

For instance, risk assessors will carefully work out how much dioxin (2,3,7,8-TCDD) can be dumped into the Columbia River on any given day, as if the river were in pristine condition and had not already been affected by dioxin and/or other persistent, synthetic, toxic, chlorinated chemicals (e.g., the pesticide DDT, and polychlorinated biphenyls from fluids used in electrical transformers) in the water and sediments for decades; as if Hanford Nuclear Reservation has not been releasing radioactive substances and toxic chemicals into the river for half a century; as if dozens of industries had not been dumping toxic materials into the river for decades (see figure 6.1); and as if the habitat has not been altered radically by the building of fourteen dams on the river.

When you ask these assessors what effect their recommended amount of dioxin will have on a Native American or Asian American resident whose immune system is poorly functioning and who eats much more Columbia River fish than the average person, they just look at you blankly. They know they haven't taken such a person into account, that their formulas and averages don't accommodate such a person, and that therefore they aren't really talking about the real world.

The most they answer (when they answer at all) is that all the other assumptions in their risk assessments are "conservative." By this the assessors imply that the numbers they have put into their formulas assume "unrealistically" large exposures and "unrealistically" high toxicity. In this way they claim that even if they haven't considered all types of stresses (e.g., high consumption of toxic-chemical-contaminated fish) they have surely overconsidered some of the stresses. At that point, they have left science, and you are left arguing with numbers that might as well have been pulled out of thin air.

There are at least four main realities that generally make it impossible for scientists to claim a hazardous activity or substance is "safe" or will cause "insignificant effects":

1. A hazardous activity or substance may cause many different adverse effects.

A risk assessment pronouncing a hazardous activity "safe" or "insignificant" will not have considered all potential impacts of that activity. How many risk assessments of a pesticide, for instance, consider all possible effects on the immune system or the nervous system, plus effects on the sexual hormone system of a developing embryo, plus all possible interactions with chromosomes and DNA, plus effects on the emotional states of humans, plus effects on vision and hearing, plus effects on the intellectual development of embryos and infants? Answer: None.

Suppose the risk assessment of a particular toxic chemical considers 40 possible adverse effects (called "endpoints") but doesn't consider whether the chemical adversely affects the brain development of human embryos. Then suppose damage to embryonic brain development is one consequence of that particular toxic chemical. The risk assessors, having not looked at this "endpoint" of health damage, will have falsely pronounced the chemical "safe."

Likewise, while ecological risk assessments of cattle and domestic sheep grazing on arid western lands have often discussed how livestock (1) frequently destroy stream banks; (2) cause sedimentation of stream-bed gravels needed by certain fish for laying their eggs; and (3) remain too long in one place, eating too much of the vegetation, very few of these risk assessments have noted that herds of these domesticated animals also destroy microbiotic crusts. These crusts are thin, fragile, slow-growing mats of blue-green algae, lichens, moss, and fungi on the surface of the soil of arid lands (Johansen 1993). These rather insignificant-looking crusts are important ecologically in desert areas because the blue-green algae's filaments hold soil particles in place and absorb rainwater. The filaments swell up in the rain, further hold the soil in place, and retain water for other plants. Blue-green algae in the microbiotic crust pull nitrogen out of the air and feed it to the larger plants, which could otherwise not obtain that nitrogen (Johnston 1997). In many arid portions of western North America, plant growth is limited by inadequate stores of nitrogen. In other words, microbiotic crusts play an important, long-term role in the ecology of arid land, and yet, for the most part, risk assessments of livestock grazing pay no attention to this factor at all.

Polychlorinated biphenyls were used for many years to prevent U.S.-made electrical transformers from overheating. Whenever these transformers exploded or leaked while in service, or leaked in dumps (landfills) where transformers had been discarded, PCBs entered the environment. Like many organochlorines, PCBs are extremely hard to break down and break apart, so they persist in the environment and in the bodies of animals for a long time as toxic pollutants. Also, like many other organochlorines, PCBs become concentrated in fats of animals (e.g., seals) that have consumed other animals (e.g., fish) that were also contaminated with PCBs in their fats. A female mammal passes PCBs to her embryos and newborns through her umbilical cord and breast milk (which is rich in fats). One of the critical adverse effects of PCBs is damage to the developing nervous systems of embryos. This damage later affects the young animals' and children's ability to think and tolerate frustration (Colborn et al. 1993; Daly 1990; Fein et al. 1984; Jacobson et al. 1990; Rogan and Miller 1989).

If the input of PCBs into humans were to somehow magically stop now, the PCBs already present in the body of a U.S. woman (e.g., a 20-year-old Michigan woman) would be transmitted in measurable concentrations through the placenta and breast milk to embryos for the next five generations—that is, from the 20-year-old to her great-great-great-grandchildren (Swain 1988).

When these persistent PCBs escape transformers, they make their way to the ocean via their travels through the ground to groundwater and streams or through the air after evaporating. Once in the ocean, PCBs enter the bodies of fish and fatty sea mammals such as seals that eat marine prey contaminated with PCBs. Eskimo (i.e., Inuit) women living in the remote Arctic Quebec consume fatty sea mammals and fish. As a result, the breast milk of these women contains 3–10 times the PCB toxicity of the breast milk of white women in Quebec (Dewailly et al. 1994).

No risk assessment of PCBs by any PCB-related industry or any government agency permitting the use of PCBs south of the Arctic has included estimates of damage to the central nervous system of Eskimo children living in the Arctic.

As the above examples show, no scientist can accurately say that a hazardous activity is safe, because all consequences cannot have been considered.

2. The adverse effects of a hazardous substance or activity are in addition to effects from other hazardous substances or activities.

Ours is a world of cumulative effects. Any individual (wildlife included) is subjected to multiple adverse effects simultaneously. Some of these effects are experienced for years.

Think of yourself. You have been contaminated with toxic and, probably, radioactive pollution through polluted air, water, and food. You have probably experienced loss of open spaces you once had for relaxation. You have probably experienced motorized noise for days on end. The ozone layer has been thinned above you, exposing you to UV-B radiation. You were exposed to toxic substances when you were developing as an embryo, and some may have damaged your genetic system. And so on.

Risk assessments that pronounce a hazardous substance or activity safe or insignificant cannot consider all relevant cumulative impacts you or any other living being has individually experienced.

Dioxin (2,3,7,8-TCDD) and dioxin-like compounds (i.e., certain other dioxins and furans, and other chemicals that behave like dioxin in our bodies) adversely affect the immune system, the endocrine (hormone) system, and the nervous system, and help cause cancer. Almost all dioxin and dioxin-like compounds have been introduced into our world since the 1920s by industrial uses of chlorine. The U.S. Environmental Protection Agency notes that "in all studies, both in laboratory animals and in humans, incremental exposures are being added onto an existing body burden that is present at birth and appears to increase with age" (USEPA 1994b). This "existing body burden" is human-caused.

Any other impacts (from whatever source) to our immune system, endocrine system, central nervous system, and tissues susceptible to cancer are being added to the effects of our "existing body burden of dioxin."

What these two examples show is that estimates of "safe" exposure to a particular hazardous substance or activity are indefensible because they pretend that we are not already damaged, or contaminated with dioxin, or otherwise under stress. Risk-assessment scientists assume in their calculations that the activity they are analyzing is being experienced by a healthy, uncontaminated, unstressed organism living in a healthy, unpolluted world. These scientists or risk assessors are pretending the world doesn't exist.

3. Organisms have differing inherited abilities to cope and unique histories of exposure to hazards.

Each of us carries an array of genes, some of which confer unusual susceptibility to certain types of damage, such as breast cancer. Others among us have inherited a remarkable genetic ability to fight off particular diseases, such as breast cancer. We are born with immune systems that vary vastly in competence.

Later, our life experiences, for instance a job we have, might vastly alter what we were born with. Exposure to pesticides while picking tomatoes, for instance, might reduce our ability to tolerate other synthetic chemicals (Ashford and Miller 1990a,b). Someone who has become chemically sensitive as a result of exposure to pesticides might become unable, for instance, to tolerate the exhaust of cars on a city street.

Certain plants could likewise be particularly vulnerable. A particular Utah canyon, for instance, might contain a small population of a plant species that exists only in that area of Utah. The existence of the entire species, then, may be threatened by cattle grazing in that canyon.

The point of these examples is that people and other animals and plants vary a lot when they begin life, and get variously damaged as life goes on. These individualistic variations in genes and life histories cannot be accounted for in a risk assessment. What may be "safe" for one individual in a species may be highly dangerous for another individual of that same species.

4. We don't understand all indirect and interrelated consequences within our complex environment.

Since there is no way we can know or predict the multiple, interdependent effects of cumulative stresses, or even all the effects of individual stresses, we repeatedly encounter "new" effects and potencies we had not expected. In recent years, for instance, we have been surprised by the following:

• the complexity and power of acquired immune deficiency syndrome (AIDS)
• sexual hormone disruption throughout the animal kingdom (e.g., turtles, fish, birds, mammals)
• the growing resistance of humans to certain life-saving antibiotics
• interactions and interdependency among the immune system, the hormone system, and the nervous system
• rising rates of breast cancer
• ozone depletion by methyl bromide, not just CFCs
• high rates of tumors and deformities in fish in certain lakes and rivers
• the apparent worldwide disappearance of frogs.

At one level, we should expect something like a rise in immune-system diseases. For instance, we know from tests that many chemicals cause changes in the immune system. We know that organisms are being exposed to immune-suppressing synthetic chemicals, to UV-B radiation (through the thinning ozone layer), and to pharmaceutical drugs that are intended to suppress the immune system. We find immune-suppressing synthetic chemicals such as dioxin stored in the bodies of humans and other animals. We admit that we have enormous gaps in our understanding of how the immune system responds to toxic chemicals (Luster and Rosenthal 1993).

But even so, we somehow seem surprised when people or ecosystems actually get sick or break down. We are surprised by the AIDS epidemic. We are surprised by rising rates of prostate cancer. We are surprised that entire frog species are dying out. Risk assessments didn't predict these problems.

Why didn't they predict the AIDS epidemic, or disappearance of frogs? Because these poorly understood, unsuspected manifestations cannot be predicted with any specificity, so they cannot be included in a risk assessment that is seeking to establish a threshold of "safety" or "insignificance" for some specific hazardous activity or substance.

These examples show that scientists do not understand all the linkages among living and non-living systems in the world and therefore cannot accurately say that a particular hazardous activity or substance will be "safe."

When "Safe" Means "Dangerous"

Despite all of the above reasons why scientists cannot determine safety through risk assessment, the term "safe" is regularly used by business and government representatives, by politicians, and even by representatives of environmental groups.

For instance, on August 3, 1996, President Clinton commented in a radio address on the 1996 Food Quality Protection Act (P.L. 104-170), which he was about to sign. This legislation had been recently passed by Congress as an amendment to the Federal Insecticide, Fungicide and Rodenticide Act and the Federal Food, Drug and Cosmetic Act. Among other provisions, this law requires that "available information" on children's consumption patterns and special susceptibilities to a given pesticide be considered when determining how much pesticide residue can remain on foods they eat. "I like to think of [this law] as the 'peace of mind' act," Clinton stated, "because it will give parents the peace of mind that comes from knowing that fruits, vegetables and grains they set down in front of their children are safe." (Clinton 1996)

Neither more realistic risk assessments nor reductions in pesticide residues will make the remaining pesticide mixtures on infants' and children's food "safe."

Representatives of environmental groups are most likely to use the word "safe" (or a variant of that word) when describing improved practices that they are proposing (for instance, organic farming that uses only biological pesticides, or a more stringent limit on a factory's toxic-waste discharges to a river), or even as a goal for proposed legislation. However, these groups should not give legitimacy to this word in connection with such human activities. Accuracy will be better served if the word "safer" or the phrase "less hazardous," "less harmful," or "more protective" is used.

Many in the public have a strong emotional attachment to "safe" as a word or concept, so it is tempting to use the word in connection with campaigns to improve human behavior. However, generally the word "safe" is not defensible. "Safer" or "less hazardous" appropriately connotes an understanding that we humans can have destructive effects on many webs of life, in ways we have not even begun to fathom. We can only try to be more careful of how we treat the world and each other; we can only try to walk more lightly on the Earth than we are now walking.

5

When Decision Makers Become Compromised: Pronouncing Unnecessary Hazardous Activities "Acceptable"

Decisions based on risk assessment involve a moral dimension when they result in subjecting others to unnecessary risks to which they do not consent.

Writing Permits to Cause Damage

A preponderance of the ways our environmental and public health are degraded results from activities that Congress has authorized through legislation and government permits. Many of these permits are based on risk assessments. Four examples of harmful results from permitted activities follow.

Contamination of Water, Air, Soil, Food Chain, Wildlife, and Our Bodies with Poisons

Industrial facilities and businesses receive permits to produce, discharge, and "dispose of" toxic chemicals into communities and ecosystems; citizens are permitted to dispose of their bodily wastes into water; wealthy communities and countries receive permits to dispose of their toxic wastes in poor communities, often minority, and in poorer countries (see, e.g., Bryant and Mohai 1992). Farmers and public agencies receive permits to release toxic pesticides out of airplanes, hoses, and underground shafts.

Elimination of Scarce But Critical Wildlife Habitat—e.g., Wetlands, Ancient Forests, and Riparian Areas[1]

Developers are permitted to drain wetlands, the timber industry is licensed to remove ancient forests from public lands, and the livestock industry is given decades-long permits to run livestock in desert streams. Communities and industry obtain permits to build and operate industrial operations on the banks of rivers and streams.

Rapid Extinction of Species

We are aware that activities that we are permitting are likely to drive certain species off the Earth.

For example, the Pacific Northwest once was home to salmon species, subspecies, and races, each individual of which knew a particular river or stream as the home in which it had been hatched, from which it had traveled to live and mature in the ocean, and to which, at the end of its life, it returned to lay or fertilize eggs and die. Our society has systematically permitted the destruction of the salmon since the early 1800s. The government has licensed dam construction and operations that block the travels of the young salmon downstream and the old salmon upstream. It has permitted logging, river-bottom road construction, and livestock grazing that break river banks and denude the land next to streams, allowing soil to wash into the stream bed, which prevents salmon from laying or fertilizing their eggs and smothers eggs already laid. These same activities allow sunlight to reach the stream directly, heating the water to temperatures lethal to developing salmon. Our society has permitted overfishing of salmon for commercial purposes.

Our society has responded to depleted salmon runs by licensing the "halfway" technology of fish hatcheries, which have attempted to grow more salmon without restoring the cause of the salmon decline (i.e., degraded habitat).

As one fish researcher points out, the fish hatchery approach will fail for at least six reasons (Meffe 1992):

• Data show that even after decades of hatchery production, salmon continue to decline.
• Hatcheries are expensive and divert resources from habitat restoration.
• Hatcheries are not sustainable, because they require continual input of money and energy.
• Because of the lack of genetic diversity in hatchery-bred fish, hatcheries introduce genetically maladapted, non-wild fish.
• By boosting the total number of salmon in the ocean and rivers, hatchery production leads to increased harvest of declining wild populations.

• Hatcheries conceal from the public the truth of real salmon decline.

Another example: Both public agencies and private ranchers have systematically permitted the poisoning and shooting of prairie dogs on public and private lands, on the supposition that these native animals "compete" with cattle for grasses and forbs. Farmers and ranchers have been permitted to plow and cultivate prairie-dog habitat for agricultural production. But the black-footed ferret feeds only on prairie dogs, and so the activities that reduced the numbers of prairie dogs also reduced populations of the black-footed ferret. Eventually, only eighteen individual black-footed ferrets were known to remain.

The Fish and Wildlife Service now is undertaking an extensive breeding and recovery program for the black-footed ferret. However, the Forest Service, the Bureau of Land Management, and state and local government agencies continue to permit poisoning and shooting (including recreational shooting) of prairie dogs and off-road-vehicle use, livestock grazing, and oil and gas development in prairie-dog habitat. In other words, our society is continuing to permit the same human activities that caused the near extinction of the ferret and/or that threaten the success of efforts to restore ferret numbers (Predator Project 1993).

Rising Rates of Breast Cancer

We have even issued permits for activities that have caused breast-cancer rates to rise. The medical establishment has been permitted to use large x-ray doses for mammograms and other diagnostic purposes (Gofman and O'Connor 1985), and these appear to have been a major cause of the current increased rates of breast cancer (Gofman 1995). At the same time, industry and agriculture have been permitted to pollute air, water, and food with hormone-disrupting chemicals that find their way into the breasts of women throughout the world (Thomas and Colborn 1992). Some proportion of breast cancer appears to be a function of disruption of hormones in the breast (Davis 1993).

Clearly, some current environmental degradation is not, strictly speaking, permitted. For instance, some toxic wastes are illegally dumped, some industries exceed their permits for discharges of toxic chemicals, and some people poach rare animals (although many of those animals had become rare because of permitted activities).

In addition, human numbers are now extremely large. Sometimes human activities that would be only minimally hazardous to the environment if limited numbers of humans were engaging in them (e.g., hunting, fishing, waste disposal, agriculture, fires, home construction, road building) become destructive when large numbers of people are engaging in them in a small area.

To the extent that governments and others encourage population growth, they support a major source of environmental impacts. As Daily and Ehrlich (1992) point out, the impact of any human population is a function of its production and consumption of resources (affluence), its technology for producing and disposing of materials, and its numbers. Any one of those three sources of impacts can itself cause adverse environmental impacts. The United States supports population growth with a tax break of $2000 for each child in a family (Internal Revenue Code, § 151(d)(1) 1994. 6 U.S.C. §§ 151, 152, 1994). In another example, Wild Oats Markets, a natural-foods chain explicitly committed to supporting the environment, gives a monetary "baby benefit" to workers. This chain's 1996 staff handbook notes: "It is important that Wild Oats staff members continue to reproduce themselves so that we will have new managers for the next century! To encourage that, we will give you $200 should you become a new father or mother."

The Moral Dimension

The social process of permitting unnecessary destruction of the environment and public health is rationalized in the modern world through the process of risk assessment, which pronounces so-called private or even public hazardous activities "acceptable," "safe," or of "insignificant" consequence and then permits them.

Of course, the concept of "acceptability" of hazardous activities is ultimately underlain, at least implicitly, by the assumption that the hazardous activities are also beneficial. For instance, the acceptability of driving salmon species to extinction with dams is underlain by an assumption of benefits—e.g., industrial growth, profits for the aluminum industry,[2] public electricity (White 1995). The acceptability of causing high rates of asthma through dust, incinerator particles, and toxic air emissions is underlain by the assumed benefits of motorized vehicles and "profitable" industries. the stability of local communities is often used as a justification for logging the remaining stands of old-growth forest in the Pacific Northwest.

There are, of course, numerous reasons to question the reality or strength of the ultimate, often thoughtlessly assumed, "benefits" justifications for a hazardous activity. For instance, those who bear direct hazards of a given activity may obtain no direct or indirect benefits from the activity. The hazards (for instance, slightly lower IQ, slightly lower sperm counts) may not be directly comparable to the benefits (for instance, profits to Monsanto, "jobs"). Many of the benefits may be available through an alternative activity that is not as hazardous.

Nevertheless, permitting of dangerous activities has always been ultimately based on the assumption or the supposition that someone benefits or from those activities.

Passing a hazardous activity through a formal risk assessment, pronouncing the estimated impacts "acceptable" on some supposedly "objective" or "scientific" basis, and then permitting the activity is a more recent procedure.[3] Alternatively, sometimes an activity is permitted without having been passed through formal risk assessment. Later, if the activity is challenged because it is proving dangerous to someone or some habitat, the activity may then be assessed and pronounced "acceptable" or "safe" via a risk assessment.

This process of permitting dangerous activities on the basis of risk assessment means that someone is judging for others that the damage those others will suffer at the hands of someone is objectively acceptable. While permitting of hazardous activities is unavoidable to some degree in a representative democracy of 250 million citizens, we must look at the degree to which communities are requiring and allowing government and business to pronounce the adverse effects of unnecessary hazardous activities acceptable when in fact the victims may not find them acceptable at all.

Labeling human activities "acceptable" when the activities are both harmful to humans or wildlife and unnecessary seems immoral to some people. By two of the principles stated in chapter 1 of this book, it is not acceptable: It is not all right to damage or kill people when there are reasonable alternatives, and it is not all right to damage or kill those who have no human voice (e.g., birds, wolves) when there are reasonable alternatives.

(Again, for a variety of reasons, I am not including wildlife-respecting hunting in this concern.)

In a rather remarkable presentation of the moral problems of permitting hazardous activities, Merrell and Van Strum (1990) posed the question of what would happen if our pesticide risk assessments were so scientifically advanced that assessors knew not only exactly how many people would be harmed by a particular pesticide but also which individuals would be harmed. A permit to release a carcinogenic pesticide into the food system, for instance, would be preceded by a list of who would contract the cancer. This, however, would constitute premeditated murder. The marked people would be entitled to an injunction on using that pesticide by their Constitutional right to life. However, bureaucrats and the private sector routinely "get away" with this premeditated murder because the victims are individually anonymous. Merrell and Van Strum write:

. . . any moral demarcation at all between pesticide "negligible risk" executions and the Nazi gas chambers of Auschwitz-Birkenau is extremely thin and must be vacillating uncontrollably. In both situations, a perceived greater public good—and lack of effective public opposition—justify the sacrifice of innocent human lives. Furthermore, the cumulative load of harmful chemicals to which humans are exposed suggests that there may be little difference in the number of victims.

The Way to Make Moral Decisions

As was noted above, it is impossible to avoid making at least some such compromised decisions. It is impossible to undertake most business or government activities without posing some hazard to workers, the community, wildlife, or the environment. In a democracy, some government employees will inevitably be required by law to make decisions about what activities will be permitted.

Within this necessarily compromised context, moral decision making would appear to require giving priority to least-harm alternatives that are most beneficial for the environment and for the public interest. However, current laws sometimes preclude consideration of alternatives when decisions are being made about hazardous activities. For instance, the Federal Insecticide, Fungicide and Rodenticide Act precludes consideration of alternatives (e.g., that more benign alternatives exist) when registering a pesticide for sale and use in the United States (7 U.S.C.A., Section 136a(c)(5)):

The [EPA] Administrator shall not make any lack of essentiality a criterion for denying registration of any pesticide. Where two pesticides meet the requirements of this paragraph [i.e., its composition warrants claims made for it,[4] its labeling is in order, and it will not generally cause unreasonable adverse effects on the environment in light of economic benefits], one should not be registered in preference to the other.

In other words, registration of a carcinogenic pesticide cannot legally be denied on the basis that less murderous pesticides exist. Thus, the national pesticide law authorizes premeditated murder, because carcinogenic pesticides must be registered as long as their benefits (i.e., profits to some people from using them) outweigh their costs (e.g., cancer for some people exposed to the pesticides) (7 U.S.C.A., Section 136(bb)).

Government bureaucrats who are "required" by such laws to make immoral decisions and all those who believe that others deserve to be treated with the greatest possible respect must rail against such laws, work to change them, and inform the public why such laws should be changed. Government bureaucrats whose activities are governed by laws that require them to sanction unnecessarily harmful activities should employ all applicable laws that do allow them to require public consideration of least-harm and beneficial alternatives. No-alternative laws allow and direct government employees to act immorally toward the environment and their fellow citizens and allow businesses to behave immorally under cover of being "legal."

There is a case to be made that the U.S. Constitution in fact requires that least-harm alternatives be guaranteed by the government. Vic Sher, a former president of the Sierra Club Legal Defense Fund, has written of requirements based on both constitutional and civil rights to do alternatives assessment (Sher 1987). Regarding constitutional rights, he notes that both federal and state courts have recognized the right to freedom from harm to our bodies as "a basic and constitutionally predicated right" (*Bouvia v. Superior Court*, 179 Cal. App.3d 1127, 1139, 1986).

Sher compares the rights of citizens not to have their bodies invaded with toxic chemicals (e.g., via state aerial pesticide programs or industry pollution permits) to court rulings supporting the rights of mental patients to refuse anti-psychotic drugs. In one such case (*Rogers v. Okin*, 634 F.2d at 653) the court noted that "a person has a constitutionally protected interest in being left free by the state to decide for himself whether to submit to the serious and potentially harmful medical treatment that is represented by the administration of anti-psychotic drugs." The same court stated that the existence of the right to be free from such intrusion is an "intuitively obvious proposition" arising out of the "penumbral right to privacy, bodily integrity, or personal security."

Another court wrote the following in rejecting a state plan to apply anti-psychotic drugs (*Bee v. Greaves*, 744 F.2d 1387, 1396 (10th Cir. 1984), cert denied, 105S.Ct. 1187, 1986):

In view of the severe effects of anti-psychotic drugs, forcible medication cannot be viewed as a reasonable response to a safety or security threat if there exist "less drastic means for achieving the same basic purpose." . . . Our constitutional jurisprudence long has held that where a state interest conflicts with fundamental personal liberties, the means by which that interest is promoted must be carefully selected so as to result in the minimum possible infringement of protected rights. . . . Thus, less restrictive alternatives . . . should be ruled out before resorting to anti-psychotic drugs.

If U.S. courts say that mental hospitals must consider less drastic means than forcible administration of drugs to ostensibly help a person, why shouldn't users of toxic chemicals have to consider alternatives to their forcible exposure of people (workers and citizens)?

Under the Civil Rights Act, courts have developed the principle that discrimination against minority groups is illegal when alternative actions are available that would not result in discrimination. For instance, the ruling in *Georgia State Conference of Branch of NAACP v. State of Georgia*, 775 F.2d 1403, 1417 (11th Cir. 1985)[5] stated that citizens who feel they are being discriminated against on a racial basis may win their case if they propose an effective alternative that eliminates that discrimination.

An example of discriminatory decision making and alternatives to discriminatory decision making may be found in permits to the pulp industry. Nine chlorine-using pulp mills are permitted under EPA standards to discharge dioxin and other organochlorines into the Columbia River. Many members of Indian tribes in the Columbia River Basin of Oregon and Washington consume six to eleven times more fish than the consumption rates upon which federal and state standards of "acceptable" dioxin discharges to the Columbia River are based (CRITFC, undated). These Indians are therefore subjected to risks higher than what is "acceptable" (McCormack and Cleverly 1990).

Chlorine compounds are used by pulp mills to make the pulp and paper white. However, technologies are available that produce white paper without using any chlorine compounds. (See chapter 6.) If the pulp mills used these technologies, they would not be discharging dioxin or other organochlorines into the Columbia River, and thus they would not be contributing to the poisoning of Indian tribes. Since alternatives exist to using chlorine compounds in pulp mills, and since use of chlorine compounds results in discrimination against Indians, the permit to discharge dioxin into the Indians' food through use of chlorine-containing compounds in pulp mills would appear to violate the Civil Rights Act.

Sher's paper and the situation of Indians of the Columbia River region point to the fundamental need to switch to alternatives assessment if we are going to respect the constitutional and civil rights of our fellow citizens (as well as avoid ecosystem degradation). The rights of Indians to nurse infants without poisoning them with dioxin in their breast milk, for instance, must surely rank higher in our society's priorities than a particular brightness of paper (USEPA 1993b).

Democracy

In addition to raising fundamental moral questions about how we treat others, risk assessment raises questions about our commitment to democratic values. Four such considerations are described below.

1. The usual risk assessment is inaccessible to laypeople.

Many government decision makers now say that communities should be involved in risk-assessment processes affecting their communities, but there is a major barrier to such participation: Most citizens cannot read a formal risk assessment, with its quantification of data, formulas, models, hidden assumptions, unstated uncertainties, selective use of scientific information, and biases.

If citizens cannot read the risk assessment, they cannot effectively participate in a public debate about the realism of the risk assessment, or whether the damage that may be caused is unacceptable to them. This is extremely dubious, since what may be at stake in the decision making is their quality of life, or their life, or the lives of others for whom they feel a responsibility, whether butterfly species or future generations of humans.

2. Risk assessment obscures and removes the fundamental right to say "no" to unnecessary poisoning of one's body and environment.

When citizens question hazardous activities on risk-assessment grounds, they are forced to operate on the playing field of formulas, models, quantification of data, hidden assumptions, biases, and selective use of information. If the citizens try to wade through this, they may forget that they have a right to deal with much more fundamental questions, such as the following:

• Is this hazardous activity essential?
• Do more sensible alternatives exist?
• Should this community have a right to say no to being poisoned?
• Does anyone have a right to take away something that belongs to us all (e.g., the Earth's ozone layer, the nation's public native grasslands, salmon species, migratory-bird flyways)?

3. Most risk assessments assume that potentially damaging behaviors are innocent until proven guilty.

In these situations, the burden of proof is placed on those who contend that the behaviors could cause harm.

For instance, if no one has studied the effects of particular mining wastes on frogs, it will be assumed that mining wastes leaching into streams will cause no harm to frogs.

Most pesticide risk assessments do not consider whether the endocrine (hormone) system of an embryo (human or wildlife) will be disrupted. By default, the pesticide is then considered harmless to hormone systems.

There is no logical reason for assuming the innocence of potentially damaging behaviors. It would be just as logical to assume, for the purposes of decision making, that a potentially damaging behavior will cause damage. With such a precautionary approach, those who would seek permits for such a behavior could be expected to consider their options for preventing all harm or causing least harm.

4. Many risk assessments use public money to justify harm to the public.

All risk assessments prepared by local, state, and federal agencies in the process of permitting hazardous activities are prepared with public money. This money from the public is ultimately turned against the public if, through risk-based permits, citizens are subsequently radiated, deprived of public wildlands, or condemned to earn their living in a workplace of chronic poisoning when alternatives that would have caused less harm to them were not considered fully in the assessments.

Quantitative risk assessments cost a lot of money. This raises the question whether it is legitimate to assess only harmful activities and exclude benign activities when the public not only is paying for the assessments but also will most likely be harmed by the permitted activities.

A California law called "California's Safe Drinking Water and Toxics Enforcement Act" (Proposition 65, 1986) requires that any business that releases a carcinogen in its workplace inform the public that it is doing so. This has led to the preparation of risk assessments to determine whether a particular chemical is a carcinogen. The cost of these risk assessments gives an idea of the cost of assessing other toxic chemicals in our society.

Cranor (1992) notes that "according to evidence presented at the [California Proposition] 65 Science Advisory Panel, it takes an agency such as the California Environmental Protection Agency from 0.5–5.0 person-years per potency assessment using conventional risk assessment procedures." Cranor presents a chart showing that the California EPA took 1250 days per chemical to assess the potency of 200 carcinogens, at a total public cost of $80 million dollars per potency assessment. This, of course, is just the cost of determining how potent a carcinogen a chemical is. It doesn't include the cost of determining whether the chemical causes birth defects, hormone disruption, liver damage, nerve damage, and so on.

If public money is going to be spent on assessments, it would seem that the public should be assured that least-harm alternatives are being assessed alongside dangerous alternatives.

How Undemocratic Can Risk Assessment Get?

Current U.S. Supreme Court Justice Stephen Breyer wrote in his 1993 book *Breaking the Vicious Circle* that he was concerned about the patchy, inconsistent way risk assessments were prepared and applied in the United States. Breyer said he was concerned about the endless arguing about many of these risk assessments, particularly by the public, who, he indicated, did not have a clue how to judge or prioritize risks accurately. He was also concerned that, because there were so many messy uncertainties in risk assessment, Congress and regulators got mired in politics when preparing risk assessments.

Justice Breyer thinks that someone who is rational should take charge of this whole process to make it orderly. The solutions he proposes in *Breaking the Vicious Circle* are worth looking at, because they are very clear articulations of how anti-democratic we would have to get in order to make people knuckle under to risk assessments.

Breyer proposes forming a small, centralized, politically insulated cadre in the Executive Branch that would rationalize risk assessment for that branch of the U.S. government. This cadre would define what amount of risk is *de minimis* (i.e., insignificant). It would prioritize federal programs on the basis of the risks the programs are addressing, and it would move allocated money from one program to another even after Congress had voted to allocate the money differently.

Within Breyer's scheme, it would be necessary to tell some people they no longer have a say in decisions about certain hazardous activities. As Breyer notes, achieving "the public's broader health and safety goals" may require "forgoing direct public control of, say, individual toxic dumps." In other words, if someone in the government decided to locate a toxic dump in a particular neighborhood and declared the toxic dump "safe," the neighborhood might not be allowed to have any say in that decision.

Breyer notes that it would also be necessary to require everyone to accept the centralized risk-assessment cadre as authority on whether something poses insignificant risk:

[The] existence of any such government-wide system would make it easier, hence more likely, for EPA's central or local administrators (say, those concerned with cleanup of wetlands in New Hampshire) to stop short of "the last 10 percent," because following a system—even one that guides rather than dictates—offers the local administrators insulation and protection from criticism. They can answer the locally posed question, "Is our swamp clean now?" with, "Yes, the swamp is clean; the risks are insignificant; and national technical (system-based) standards say that is so."

Breyer doesn't address the question of what "those concerned with cleanup of wetlands in New Hampshire" should do if they know that the "swamp" that has just been pronounced clean isn't clean. What if those concerned citizens know that the risk assessors didn't really understand or care at all about wetlands and thus mislabeled a contaminated swamp a clean one?[6]

In 1994, for instance, Lake Apopka (near Orlando, Florida) was on the brink of being declared "cleaned up" by the EPA after a 1980 spill of the pesticide dicofol by Tower Chemical. Tower Chemical's production of dicofol had involved boiling DDT in sulfuric acid. By 1994, the levels of DDT in Lake Apopka were believed to be too low for the chemical to cause cancer (Campbell 1994). Only during a scientific study of the potential for alligator farming in Lake Apopka was ongoing damage discovered: the hormone systems and the genitalia of alligators were discovered to be highly compromised, and these effects were found to be associated with the dicofol spill (Guillette et al. 1994; Colborn et al. 1996).

Breyer's vision is nightmarish and anti-democratic. In *Breaking the Vicious Circle*, he rarely strays from the topic of cancer risks caused by single chemicals. Breyer seems to be ignorant of such complexities as the multiple and indirect non-cancer impacts of toxic chemicals (e.g., reduced male sperm, reduced alligator penis size), the unknowns about mixtures of toxic chemicals, and ecosystem risk assessments. He also appears fundamentally ignorant about the value-laden nature of preparing risk assessments.

We will always need many citizens who refuse to remain passive when some bureaucrat tells them that their local contaminated wetland is clean because national technical standards devised by some distant cadre say it is clean.

Breyer's conception of risk assessment is starkly anti-democratic and unscientific. All he is really doing is revealing the fundamental values and morals that are involved in how and whether we permit unnecessary hazardous activities.

6

When a Society Isn't Serious About Environmental Health: Assessing a Narrow Range of Options

Risk assessment is not a legitimate process of assessing possible harms of activities when the only activities being assessed are harmful.

Pretending That We Can't Think of Anything Better: Four Examples

The following four examples illustrate our society's pattern of acting as if we have to cause harm, as if we have no alternatives.

Incineration of Hazardous Waste
The information in this section is taken from a well-documented Greenpeace report on hazardous-waste incineration, Playing With Fire (Costner and Thornton 1990).

Every minute, U.S. industries generate almost two million pounds of hazardous waste. The society has become aware that when we dispose of these wastes into rivers and lakes, the entire receiving aquatic system becomes polluted and damaged. Many of the wastes resist quick natural breakdown.

We are gradually facing the reality that when we dispose of these wastes in landfills and deep wells, persistent chemical pollutants eventually migrate out into groundwater, surface water, air, and food chains to poison living beings.

Some industry people and government representatives promote the belief that incinerators will take care of these wastes. This means we can throw massive quantities of diverse chemical mixtures into a really hot incinerator, and harmless gases and benign ash will come out the other end.

But the incinerator doesn't completely burn all wastes. If we manage to capture some of the resulting unburned pollutants in wet scrubbers,[1] we then partially treat these and discharge them into rivers. If we capture some of the unburned pollutants as fly ash[2] in dry pollution control devices, we then bury this polluted ash in landfills. However, all landfills leak (Montague 1992b).

To make matters worse, the incinerators create new compounds that are "more difficult to destroy and . . . more toxic than the parent compound" (USEPA 1985b). Many of these incineration-produced compounds are organochlorines, compounds that contain at least one carbon atom bonded to chlorine. Many organochlorines are extremely toxic, persistent, and bioaccumulative: Dioxins, furans, and hexachlorobenzene are examples.

Some of these "new" compounds are fragments of chemicals that were fed into the incinerator. Hexachlorobenzene,[3] for instance, is produced when PCBs are broken apart while being burned. Some incineration-produced compounds are complex recombinations of chemicals fed into the incinerator. Naphthalene is formed this way. Finally, some are simple fragments that are produced universally whenever chlorinated organic (carbon-containing) compounds are burned. Carbon tetrachloride, benzene, and chlorobenzene are examples (USEPA 1989).

And of course metals (e.g., lead, mercury, cadmium, chromium) cannot be destroyed by incineration, so the metals are emitted into the air or caught in fly ash, bottom ash, and scrubber effluents.

Then there are "upset conditions" caused by equipment failures, human errors, or sudden changes in the waste that is being shipped and fed into the incinerator. Explosions and flame-outs sometimes occur, or minor disturbances in small portions of an incinerator for brief periods of time. A portion of a hazardous-waste-burning cement kiln, for instance, may become clogged, causing backups and accidental releases of partially burned pollutants. Uneven temperature distributions, poor mixing conditions, or turbulence can likewise allow unburned chemicals to leave the incinerator.

These upsets or suboptimal conditions release unburned wastes and new products of incomplete combustion into the air. As the EPA Science Advisory Board notes (USEPA 1985b), "even relatively short-term operation of incinerators in upset conditions can greatly increase the total incinerator-emitted loadings to the environment."

Because many citizens have concerns about what will be released into their communities from a hazardous-waste incinerator, risk assessments of the "best available technology" of incineration are prepared. These risk assessments are based on (a) studies of some of the toxic effects in a few selected species of animals of a few of the individual chemicals released by the incinerators, and (b) models used to estimate how large the releases may be, and how a few species of living organisms (particularly humans) may be exposed by inhalation or through skin. Ingestion of contaminated foods, a significant route of exposure, is occasionally considered.

And then government officials and people in the incineration industry come to town. They explain to the community that the risks posed by a hazardous-waste incinerator won't be significant. They say that the risk of releasing toxic materials is low, that if toxic materials are released they will be released in such small amounts that it doesn't matter, and that if damage is caused it will be minimal or "acceptable."

Concerned citizens must then marshal evidence that this optimistic scenario is not accurate. But the central question that should be on the table is whether hazardous-waste incineration should be happening at all. Does the community have to wade through a fiery stream of toxic chemicals, or is there a bridge upstream?[4]

There are at least three major alternatives to the incineration of hazardous wastes:

• If so many toxic chemicals were not used in industrial and agricultural production, only small amounts, if any, of hazardous waste would be produced.

• Cleanup technologies that do not rely on incineration, such as bioremediation (using bacteria to "eat" hazardous waste) and chemical treatments, could be used. Development of such technologies can be focused on those hazardous wastes that have already been released into the environment. In addition, development of these technologies can be focused on wastes that are produced during manufacture of products widely regarded by the society as essential.

• Above-ground, on-site storage of wastes could be required. A company or a government facility has a great incentive to reduce its use of toxic substances and production of toxic wastes if it has to retain all its untreatable or non-reusable wastes on its own property. The wastes could be contained in impermeable above-ground containers that could be easily monitored for any leaks until technologies for treating or reusing the wastes were developed.

The last of these options reminds us that we need to look not only at developing technological alternatives regarding use and disposal of toxic materials, but also at developing incentives for reducing the use of toxics. In some cases low-impact technologies already exist, but incentives to use and develop such technologies are lacking.

The following examples illustrate possible incentives:

• A state could withhold a permit from any company seeking to incinerate any of its hazardous wastes unless it has developed (through a public review process) a plan for reducing its toxic chemicals use by 50 percent in 5 years, at which time it must issue a plan for reducing its toxic use by another 50 percent within the next 5 years. More than 10 years ago, the U.S. Office of Technology Assessment indicated such a reduction was feasible with technology then available and suggested that a 10 percent annual reduction in waste over a period of 5 years "appears feasible, based on reports of recent successful efforts" (OTA 1986).

• Each company bringing hazardous waste to an incinerator (or to a "chem-fuel" mixing facility, which "blends" toxic chemical brews for incinerators) would pay a 100 percent tax, part of which would go to an independent institute for the purpose of developing technologies that reduce use of those specific toxic substances that are being brought to the incinerator, part of which would go to an institute that would research the existence of least-toxic alternative processes and products available in the world, and part of which would go to non-profit organizations that inform the public about alternatives to the purchase and consumption of specific products that "depend" on use of the toxic chemicals in question.

We clearly have options to permitting harms caused by hazardous-waste incineration. We can do better than assemble a team of risk assessors that determine why a community or a region shouldn't worry about the presence of a hazardous-waste incinerator.

Making Paper with Chlorine Compounds vs. Making Paper with Wood

We are making blinding-white paper using chlorine or chlorine-dioxide at approximately 102 pulp mills in the United States and using trees from forests or plantations to make paper at all 325 paper mills in the United States (Miller-Freeman 1994). However, we don't have to use chlorine compounds or trees.

A moderate-size, chlorine-using pulp mill produces 1000 tons of pulp a day. Such a pulp mill also daily releases tens of tons of organochlorine chemicals into air, water, sludge, and its paper products (Bonsor, McCubbin, and Sprague 1988). Many of the chlorinated paper products and some of the sludge are eventually incinerated, producing additional toxic chlorinated poisons.

There are alternatives. Paper mills in Sweden were the first to show the world that white, totally chlorine-free (TCF) paper could be produced profitably. More than 30 pulp mills produce white paper without any chlorine compounds (personal communication, Albert Hattis, Director of Administration, Chlorine-Free Products Association, January 31, 1996). Only two of these are in the United States: Louisiana-Pacific in California and Lyon Falls in New York.[5] The Swedish pulp mills ended the myth that production of white paper requires chlorine technology and pollution.

The recent effort by the EPA to draw up guidelines for pulp and paper mills avoids requiring such processes for kraft pulp and paper mills[6] (the majority of chlorine-using pulp and paper mills in the country) because the paper would only be white (ISO 75-80, which is high on an internationally recognized scale of whiteness), not blinding white (ISO 90) (USEPA 1993a).[7]

In this case, the most toxic contaminant known (i.e., dioxin) is being produced by one of the most unnecessary industrial processes imaginable (making paper blinding white with chlorine or chlorine dioxide). The EPA's stated excuse for failing to require TCF pulp-mill processes is that the paper would not be white enough to meet ISO 90 (USEPA 1993a; see p. 66109).

According to a recent report on alternatives to using wood (Soltani and Whitney 1995), only 10 percent of the inhabited Earth is roadless. More than 350,000 miles of logging roads have been cut through forests in the U.S, which is more than seven times the length of the U.S. interstate highway system. Though 95 percent of the ancient forests of the United States are gone, the U.S. logging industry expects a 46 percent increase in logging operations by 2040. Some of that would come from the remaining old-growth forests.

More than 80 percent of the trees cut in the United States are reduced to building materials and paper. On the other hand, 300 mills around the world are making paper without wood. They are making it from rice and barley straw (China), sugar cane waste (Mexico and India), and hemp (France). Two other wood substitutes that are promising for the United States are straw (the leftover stalks from cereal grain production) and the kenaf plant. A small mill in British Columbia is profitably making paper from agricultural waste. The small scale of this mill keeps capital costs low and allows it to be located near the agricultural sources. Three more such mills are planned.

The goal of the Rainforest Action Network is to reduce the use of wood in the United States by 75 percent in 10 years. While alternate building materials would play a role in this (e.g., recycled wood, earth materials, and straw bales), production of paper from wood substitutes would also reduce the use of wood (Soltani and Whitney 1995).

The existence of commercially sound alternatives to conventional paper production shows the narrowness of the alternatives we consider while we expend major effort on assessing and debating their risks.

Growing Crops in Soil Fumigated with Methyl Bromide

Methyl bromide, a fumigant pesticide, is a potent depleter of the ozone layer, and it is highly toxic, capable of causing severe neurological damage or even death to farm workers or fumigation workers exposed to it.[8]

The major use (about 70 percent) of methyl bromide worldwide is as a soil fumigant (a pesticide in vapor form), to kill all organisms in soil (e.g., nematodes,[9] weeds, insects, bacteria, fungi, and viruses), generally in nurseries or when certain seedlings, fruit trees, and vines are planted. The largest user of methyl bromide in the world (49 percent) is the United States (USDA 1993). Five crops account for 80 percent of U.S. methyl bromide soil fumigation (USDA 1993): tomatoes (primarily in Florida), strawberries (primarily in California), peppers, ornamentals/nurseries, and tobacco.

Although methyl bromide is currently slated for phaseout in 2001 under the Clean Air Act because it is a potent ozone depleter, some farmers are trying to retain its use. These farmers feel they must keep using methyl bromide because most other soil fumigants have already been banned or are highly restricted because they contaminate groundwater and poison workers.

However, there is one major alternative to using one environmentally untenable fumigant after another, and it has been used by many farmers for the last 4000 years. It is to live within the limits of biologically diverse, living soil, rather than destroying all life in the soil. Such cooperation with the soil forms the basis of a number of current agricultural practices (MBTOC 1995):

• Crops can be rotated. That is, alternating crops can be grown on a piece of land so that enemies of one crop don't build up as much as they do when that one crop is repeatedly grown on a piece of land. For instance, oilseed rape, when grown in rotation with sugar beets and potatoes, produces compounds that are toxic to certain soil fungi and nematodes that attack the beets and the potatoes.

• A piece of land can be allowed to lie fallow (unplanted) for a period of time, so that pest organisms that depend on the crop have no food and are thereby reduced in numbers.

• Organic amendments or mulches of plant material can be used, so that the soil is rich in a diversity of organisms, some of which help control organisms that are troublesome to a particular crop. Organic amendments include livestock manures, shells from shrimp and other seafood, sewage, and wood wastes.

• A crop can be planted when conditions are not favorable for the crop's "pests" in the soil. For instance, root-knot nematodes have been controlled in Georgia by a combination of crop rotation and early planting to avoid periods optimal for nematode development.

• Flooding is particularly effective against pest development when organic matter has been incorporated into the soil. As microbes digest the organic matter in the near absence of oxygen, they release compounds that are toxic to many soil-borne pests. Flooding has been effective in controlling a disease called verticillium wilt in cotton.

• "Cover crops" help to suppress problematic organisms in the soil. Cover crops are non-commercial crops that are turned back into the soil as green or dry residues. In Florida, for instance, winter vegetable production may be preceded by summer cover cropping with sorghum or Sudan grass.

• Plants can be bred to resist attacks by particular organisms, and non-resistant plants can be grafted onto resistant rootstocks.

• Biological control (i.e., ensuring the presence of living organisms that prey on or kill problematic organisms) can be used. Some soils, for instance, contain organisms that suppress certain soil-borne pests. "Beneficial" organisms from these soils can be isolated, reared, and transferred into soils that have those pests.

• Residues of certain crops (e.g., broccoli and cabbage) can be tilled into the soil. As they decompose, these residues release chemicals that are toxic to certain soil-borne pests. This is called "biofumigating."

The U.S. Department of Agriculture claims that the economic consequences of phasing out methyl bromide would be devastating (USDA
1993). Because the USDA does not focus its research on alternatives to pesticides, it assumes that no alternatives to methyl bromide exist (Knipling
1997).

There is no question that cooperating with life in soil sometimes put limits on what agribusiness might "want" to do in agriculture. For instance,
growing tomatoes in monoculture (single-crop agriculture) in the warm,
wet, pest-friendly soil of southern Florida requires enormous amounts of
numerous pesticides (USDA 1997a). Likewise, growing strawberries constantly on one piece of California land requires methyl bromide to kill
strawberry pests.

In other words, if certain crops are to be grown continuously on certain
pieces of land in certain areas, there may be no alternative to using methyl
bromide. However, saving the ozone layer is more important than growing
those luxury crops year-round in those areas. In such cases, it is reasonable
to look at alternative supports for farmers who are willing to rotate a highly profitable crop with a less profitable crop so that large amounts of pesticides will not be required. It is reasonable to determine that, if certain
crops cannot be grown in an environmentally sound manner on one particular site, maybe those crops should not be grown on that site. Sometimes
alternatives have to be broader and more complex than mere technological
"fixes" for growing, manufacturing, or consuming exactly what people currently want to grow, manufacture, and consume.

Flushing Our Toilets

A close look at a map of the Pacific Northwest region of the United States
reveals a large number of points where sewage is discharged into rivers. The
sources include publicly owned treatment works (POTWs) and combined
sewage overflows (CSOs). (When rains bring so much storm water into a
sewage treatment plant that holding tanks are filled, partially treated
sewage may be discharged directly into the river with the storm water.)

The maps don't show the numbers of septic tanks, which are holding
and/or leaking sewage underground near underground water supplies.

And then we go into a bathroom and deposit our feces and urine in clear,
clean water, as do hundreds of millions of other people.

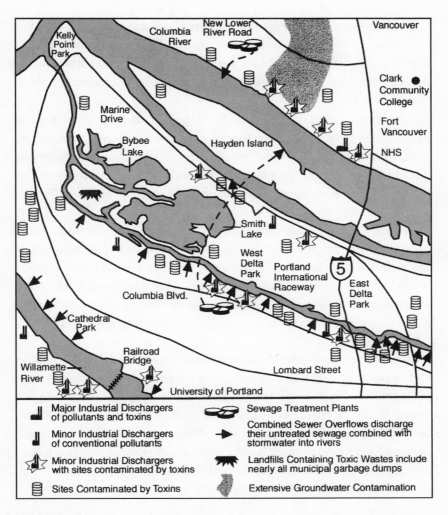

Figure 6.1
Discharge points into the Columbia River as of January 1992. Source: Northwest Environmental Advocates. Portland Vancouver Toxic Waters Map. Copyright Northwest Environmental Advocates.

This behavior has adverse environmental consequences. Sewage adds organic nutrients to rivers, which promotes heavy growth of algae and other organisms, which then consume oxygen, making the river inhospitable to the organisms that naturally live there. This process is called *eutrophication*.

Greenpeace found in 1986 that the oxygen levels in the Thames River had decreased to such an extent that agencies had to pump oxygen into the river to prevent massive fish die-offs (Rapaport 1995). The 2.4-million-acre "dead zone" at the mouth of the Mississippi was created in part by sewage discharges from 621 municipal treatment facilities along that river (Costner and Thornton 1989).

Sydney, Australia, discharges almost 5 billion gallons of primary-treated sewage[10] into the ocean per day. Massive outbreaks of red oceanic algae, called "red tides," form each summer as Sydney's sewage mixes with the warm waters of the East Australia Current. This is followed by the death of sea birds, whales, and penguins from Sydney all the way to Melbourne as the tides move south around Australia's coast in December and January (Gillmore 1993).

In addition, water-related diseases can be transmitted through contact with water contaminated with human excreta. Worldwide, these diseases kill more than 4 million children under the age of 5 per year (Engelman and LeRoy 1993). In 1991, elevated bacterial levels forced the closure of more than 2000 bays and beaches in the United States (BNA 1992). The tourism industry loses approximately $300 million per year because of beach closings, algae-clogged swimming areas, and outbreaks of cholera (Platte 1995).

Consider the following, from Costner 1990a:

Using water as a personal waste vehicle—the repository for everyone's feces and urine—is a very recent cultural phenomenon. Only a few generations have been toilet trained to water, but this experience has profoundly distorted our collective sense of reality.

Water toilets reinforce one of our mightiest taboos, the social prohibitions that surround our body products. They let us forget that feces and urine are a natural part of the food chain—repayment to the soil for the nutrients supplied in our food. Water toilets foster the illusion that these or any products can be flushed away without consequences.

. . . Where water is a waste vehicle and anything can be a waste, no water resource is safe. Our culture is permeated with this "waste/water" illusion. It is instilled in us as one of the earliest and strongest lessons of childhood—toilet training. It is reflected in the design and operation of every conventional home, city, business, factory, and farm.

There are alternatives, of course. Costner details a variety of commercially available dry treatment systems, which transform human body wastes into humus which can then be returned to the soil. Costner also describes simple systems for returning bath water, laundry water, and dishwashing water to the land.

Our industrialized societies' systematic and unnecessary pollution of water begins with our own bodies, even though we have alternatives. We could be returning our nourishment to the Earth. Instead, we first appropriate clean water and the biological production of many acres to supply us with drinking water and food, then we respond to this gift by polluting Earth's clean water. Then we undertake risk assessments of how much "biological oxygen demand" (of decomposition of organic material) we can place on our rivers without strangling river organisms, what quantity of nutrients we can add to the rivers without clogging them with algae, and how much fecal bacteria we can swim among without getting sick.

7

Who Loves, Uses, or Cooperates with Risk Assessment?

Proponents of relaxed government regulations on nuclear power, industrial pollution, occupational safety and health, environmental protection, and the like will find risk assessment, insofar as they are able to interest others in it, a very fruitful contest. . . .
—Langdon Winner (1986)

Business loves risk assessment, government agencies use it, and many scientists cooperate with it. These sectors engage in risk assessment for mostly self-interested reasons, not public-interest reasons.

That industry must love risk assessment is evident from the enormous pressure that the business sector has brought on Congress to require that government agencies prepare risk assessments and cost-benefit analyses (which include risk assessments) in advance of regulatory actions.[1] It must be that risk assessments generally serve private interests.

Many government agencies prepare or use risk assessments. Some agencies prepare risk assessments because they are required by a current law; some are prepared because someone within a particular agency wants them prepared. The agencies ultimately choose the tone and content of those risk assessments, of course, and rarely prepare assessments of risk and benefits for a full range of alternatives. Risk assessments often help agencies avoid conflict, change, or protection of the public interest.

Thousands of scientists contribute scientific information to risk assessments or even prepare the risk assessments. Some of them believe that the information will make the risk assessments better, although many of them work for companies or government agencies that will sharply limit what is considered in the risk assessments. Most know that risk assessments are conglomerations of political interests, selective assumptions, and selective data, and most admit that the risk assessments do not consider cumulative impacts.

On the other hand, many people witness, read, and hear of increasing resistance to both the content and the process of risk assessment by citizens, by community groups that advocate for environmental and public health, and by traditional indigenous groups (see, e.g., Indigenous Caucus 1994 and CIBA 1994).

I contend that the common reality underlying these divided stances with respect to risk assessment is that most risk assessments protect and justify business as usual rather than serve the environment and the public health. Most risk assessments do not result in decisions that require business or government agencies to take significantly better care of the world. If most risk assessments resulted in recognition that neighborhoods, individual citizens and workers, wildlife, and the environment are being harmed, business would not be promoting risk assessment as vigorously and as enthusiastically as it does. If most risk assessments resulted in acknowledgment that government agencies should change their behavior (for example, that they should deny permits for hazardous activities to economically and politically powerful private interests), those agencies would more actively avoid preparing risk assessments.

Let us look closer at how risk assessments serve the interests of business, government agencies, and many scientists.

Business Loves Risk Assessment

1. Risk assessment is the basis of all permits and registrations for hazardous activities and products.

Through risk assessment, an industry gets significant legal protection for activities that result in contaminating communities, workers, wildlife, and the environment with toxic chemicals. Through risk assessment, industry gets protection for filling streams with sediments (via agriculture, logging, and livestock grazing), thinning the ozone layer, causing high cancer rates, avoiding cleaning up its own messes, and other Earth-damaging activities. "We're meeting all standards as we take the top off this mountain to strip mine," an executive of a mining company might say.

2. Activities currently practiced or proposed by industry are largely taken as givens in most risk assessments.

Most risk assessments are prepared simply to help determine the extent of commercial activities and where those activities will take place, not to challenge whether the activities are necessary or appropriate.

3. Risk assessment gives industry the aura of being scientific about the "safety" or "insignificance" of its activities.

"Our pesticide is registered by the EPA, and no pesticide can be registered if it causes unreasonable adverse effects," an executive of a pesticide company might well say. Of course the pesticide company is not reminding you that "unreasonable adverse effects" are defined in the federal pesticide law as involving more than biological toxicity. But in fact they are defined as "any unreasonable risk to man or the environment, taking into account the economic, social, and environmental costs and benefits of the use of any pesticide" (FIFRA 1972). In other words, a pesticide that has highly toxic or dangerous effects may not be considered as having "unreasonable adverse effects" if it brings great economic benefits to the chemical company that makes it or to the agribusiness farms that use it.

When few studies are available for a risk assessment, and when none of those studies "prove" harm, a business is able to say "No risk assessment of our activities has shown harm." The unwarranted implication is that the company's activities are safe, even though the very next study may demonstrate that they are harmful.

4. The complexity of most risk assessments and the room for debate regarding appropriate assessment assumptions allow interminable haggling.

As long as business as usual can continue while debates about a risk assessment rage, complicated risk assessments are useful to business. New regulations can be delayed. Calls to prevent pollution can be stalled as long as consensus on whether the pollution causes more harm than good hasn't been reached.

One example of this involves risk assessments of carcinogens. It takes several years to test whether a certain chemical causes cancer, and it can take considerably longer to determine the mechanism by which that chemical causes cancer. Since many chemicals used and discharged by industry have been found to be carcinogenic in laboratory animals, industry has recently been pushing to require unassailable scientific knowledge about the mechanism by which a substance causes cancer in laboratory animals before industry is required to reduce or eliminate the use of that chemical. Industry argues that a substance that causes cancer in laboratory animals may not cause cancer in humans, often on the ground that the substance may have caused cancer in laboratory animals only because the high doses damaged tissues, and the damage then allowed cancer to develop. Since humans will not be exposed to such high levels, industry argues, human tissues will not be damaged (Ames and Gold 1990).

Industry finds this line of reasoning and debate useful because another round of tests helps delay regulations that might require elimination or reduction of its use or production of carcinogens.

If risk-assessment debates had to be resolved before industry could continue using the chemicals being assessed, industry would not be interested in extending risk-assessment debates. For instance, if industry were not allowed to produce, sell, or discharge a cancer-causing substance until all risk-assessment questions were resolved, industry would not be pushing for a requirement that we understand the mechanisms by which each carcinogen acts. It is only because industry generally is permitted to produce, release, and sell toxic chemicals as long as risk-assessment debates rage over those toxic chemicals that industry is interested in debating the exact mechanisms by which a carcinogen causes cancer.

In a society where environment-affecting activities and substances are considered innocent until proven guilty, it is in the interests of industry to use risk assessments to debate endlessly about precise innocence or guilt.

5. To read or to debate a complicated risk assessment requires technical expertise.

Since the outcome of a risk assessment depends largely on the selection, arrangement, and presentation of scientific information, it is largely controllable. However, selecting, arranging, and presenting scientific information takes technical expertise. It is generally cost effective for industry to hire people with "technical expertise" to do these jobs, since meaningful regulation could be much more costly.

Consider the case of Kettleman City, California, a low-income, rural, Hispanic-American community where most of the residents did not want a new Chemical Waste Management hazardous-waste incinerator permitted. That community challenged a 3000-page risk assessment written in English with an eight-page summary in Spanish (Rice 1991).

6. Reducing "risk" and damage by writing optimistic risk assessments is far more convenient than reducing real-world damages by ceasing hazardous activities and adopting more environmentally appropriate practices.

The power of the computer to reduce risk in charts and tables is enormous. So many of the variables in a risk assessment are malleable that an assessor can fiddle with assumptions, feeding different hazard and exposure numbers into the computer until the conclusions show an acceptable level of risk.

Government Agencies Use Risk Assessment

Preparing, defending, and altering risk assessments take a lot of staff time, a lot of paper, and a lot of meetings. Some government risk assessments are required by laws, such as "cost-benefit" laws. Other risk assessments may be required by agency regulations. Still others are done because some decision makers in an agency are afraid of not doing a risk assessment, or because everyone else is doing them, or because the decision makers simply want to do a risk assessment for a given activity. Many "government" risk assessments are contracted out, at public expense, to private risk-assessment businesses and corporations.

Some employees of government agencies know that certain activities are causing harm and would like to require less harmful alternatives. These people can be frustrated by the endless rounds of risk assessment and by the isolation of an agency's risk assessors from those who know about alternative activities. Some of these employees "live with" the legally required risk assessments by trying to take into account more factors of harm, or more exposed members of society, or more elements of an ecosystem, or the input of more citizens.

Many government employees, however, appreciate the cover that risk assessment gives them to avoid requiring change. Risk assessment allows an agency to avoid political trouble. Risk assessment serves timid regulatory agencies in the following ways.

1. Risk-assessment processes allow government permitters to hide behind "rationality" and "objectivity" as they permit and allow hazardous activities that harm people and the environment.

It is not always easy for an individual to rationalize the permitting of a hazardous activity that will, for instance, reduce a species' chance of survival or result in denial of medical benefits to veterans. Risk assessment can let employees of a government agency distance themselves from the meaning of their decisions.

Many agency employees convince themselves that risk-assessment processes really are rational and objective, and that therefore their decisions really aren't causing harm. Such employees may then focus more on whether a risk assessment has been developed according to the "rules" than on whether it reflects reality.

2. Risk assessments can be manipulated endlessly, and government agencies like to have discretion.

Because risk assessments can be manipulated to approach desired outcomes, current policies can remain in place or policies accommodating predetermined management can be installed.

3. Since risk assessments involve so much uncertainty, so many assumptions, and large doses of cumulative reality, they are not reproducible, as is a scientific experiment or study. They therefore cannot be disproved.

An agency's conclusions can be criticized, and alternative assessments can be proposed, but who is going to rule that this agency's conclusions are wrong or arbitrary? When you are building with putty, whose sculpture is the correct one?

The U.S. Supreme Court has ruled, for instance, that courts must defer to the "scientific" claims of federal agencies, even if independent scientists' evidence seems more compelling (*Marsh v. ONRC* 1989). Imagine trying to challenge an agency's risk assessment in the face of this required presumption against science, evidence, and rationality.

4. Risk assessment gives specialist power to an agency in ways that alternatives assessment does not.

Not everyone is able to prepare, read, or decipher a complicated risk assessment. On the other hand, society has many people who know about alternatives to current harmful activities.

Limiting a risk assessment to one option or a few standard options and to the habits of conventional risk-assessment companies allows an agency to work primarily with business. These standard risk assessments allow an agency to avoid dealing with many other groups who know about alternatives to the proposed activity and about unquantifiable benefits and hazards of the activity and of the alternatives.

Most agency and contract risk assessors know a great deal about standard risk-assessment techniques and very little about alternative technologies or processes. If they were to depend on citizens in the society to show them alternatives, they would lose their specialized power.

5. Risk-assessment models and numbers are intimidating to citizens, so they give agencies the upper hand.

Recall the Kettleman City case, mentioned earlier in this chapter. Even a native English-speaking community would have trouble wading through and critiquing a 3000-page risk assessment.

6. By focusing the public's attention on the details of a risk assessment, agencies divert public debates from consideration of whether the assessed activity should even be taking place.

The risk-assessment process lets agencies avoid facing the responsibility of helping a community or a society to change and the responsibility of requiring a multinational corporation to change.

The business community can be very hard on a government agency that raises questions about the necessity or the appropriateness of particular business activities. Business can go to the legislature and cut an agency's budget. Business can lean on a governor, a county commissioner, or an agency director to change uncooperative personnel.

Life is generally much easier for a government agency if it can avoid requiring the rich and powerful to change.

Nothing I have said above assumes that agency employees consciously plot to gain the upper hand, bamboozle the public, gain discretion, hide behind rationality, or divert the attention of the public from alternatives. However, if risk assessment in fact makes life easier for an agency or for an agency employee in these ways, it is likely to be very hard to turn down.

Many Scientists Live with Risk Assessment

Most scientists I have spoken with who are familiar with risk assessment admit that risk assessors cannot address cumulative impacts.[2] They admit that risk assessors do not calculate risks for the most sensitive or vulnerable individuals. Most scientists admit that many risk assessments are based on selective information, arbitrary assumptions, and enormous uncertainties rather than on pure science. They admit that economic and political considerations associated with risk management often affect how risks are calculated.

Nevertheless, scientists in the United States and around the world work on thousands of risk assessments daily. They conduct research that will be plugged into risk assessments. Many accept assignments or jobs to write risk assessments on behalf of industry. They explain risk assessments to the public. They defend or attack risk assessments in court. Some agree to change risk assessments that do not please someone who has power over them. Some scientists believe in the risk assessments they have produced, or in the power of scientific information to change a risk assessment's outcome.

In what follows I describe four things that encourage many scientists to cooperate with the risk-assessment process, despite their recognition of the arbitrary and non-scientific nature of most risk assessments.

1. Risk assessment provides salaries.

Risk assessment employs many scientists in this country. Scientists by the thousands are hired to work on the various steps of risk assessments. These scientists are often called "environmental consultants."

2. Risk assessment gives some scientists and health professionals the sense that they are encouraging rational decisions by providing accurate scientific information.

Concerned scientists can push for more elements to be considered in a risk assessment, or for more realistic numbers and better research to be included. Some risk assessments are altered as scientists provide scientific evidence that more damage is occurring or might occur than was previously estimated.

However, sometimes these scientists are simply naive. They are not familiar with the larger social history of how risk assessments have been used primarily to defend hazardous activities. They may not have heard of the pressure particular scientists have come under to alter numbers that indicate damage. They may believe that most uses of risk assessment are more scientifically objective than they actually are.

The crucial question, however, is whether these scientists are providing scientific information for a reasonable range of options or only for bad options. Carefully and fully analyzing the rather small risks to a woman of wading across an icy river (see chapter 1) may be "scientific," but it is ultimately unnecessarily detrimental to the woman if the scientist is not also analyzing the alternative of crossing on an existing bridge.

3. Risk assessment helps a scientist keep distant from pain.

Risk assessment talks about statistics, not about specific people or other animals. When cancer is calculated as "risk," a scientist does not have to think about each individual cancer victim's terror. The scientist does not have to be at the bedside of a child dying of brain cancer, for example. Risk assessment allows a scientist to talk about the "risks" from consuming fish contaminated with dioxin. That way, the scientist does not have to think about the spiritual loss to a Native American of her cultural link with those fish and that river. The image of workers being poisoned by their employers can be reduced to issues of "exposure" and "Threshold Limit Values." The logging of a watershed can be called "reduction of catastrophic fire potential."

Risk assessment is abstract. It is numerical. It rarely includes photographs or personal narratives. Risk assessment allows a scientist to avoid living, as Aldo Leopold (1953) described it, "alone in a world of wounds."

In the process of preparing risk assessments, the pain of human illness, suffering wildlife, and environmental degradation is transformed into numbers and muffled.

4. Risk assessment helps some scientists avoid acknowledging that advocacy for change rather than more science is what is needed.

How essential has it been for scientists to continue to learn more and more about the devastating effects of dioxin on life, for instance? Is that what has been really needed? We knew by 1969 that dioxin-contaminated, chlorinated herbicides caused birth defects in rats and mice at extremely low doses (Bionetics Research Laboratories 1969). Do we need to know more in order to get down to the business of alternatives to the production of dioxin (i.e., alternatives to the chlorine industry)?

Many scientists believe in the social power of scientific information. They hope that the next line of research or the next devastating piece of evidence will really convince society that some activity is harmful, and that society will then change. For instance, if they can just show that dioxin disrupts the balance of male and female hormones in humans, then maybe dioxin production will be banned.

Usually, however, a society has more than enough scientific evidence about the effects of hazardous activities to warrant change. We know that mining wastes devastate aquatic life in rivers for miles and miles. We know that nuclear wastes can't be stored safely anywhere. We know that the clear cutting of forests causes landslides and chokes streams with sediments. We know that allowing livestock to graze in arid lands eliminates streams and springs upon which almost all native desert wildlife depends. Most people know that burning fossil fuels causes global warming.

Generally it is not lack of scientific information about "risks" that prevents our society from moving toward more appropriate relationships with the environment. It is entrenched political and economic power: business wants to continue doing what it is doing. It is also a lack of public discussion about alternative ways of behaving. Social change is harder than preparing another risk assessment.

8

Unnecessary Societal Triage: Comparative Risk Assessment

Prioritizing environmental problems using risk assessment implies that some are unimportant and can be ignored. A better way to approach our multitude of environmental problems would be to rank the most effective ways to give society the incentives and ability to prevent and solve all environmental problems.

In William Styron's 1979 novel *Sophie's Choice*, a mother is given a diabolical choice by Nazi bureaucrats. She is asked "Which child will you hand over to us: Your daughter or your son?" We all know that question should never have been asked.

When we, as a society, ask ourselves the comparative-risk-assessment question "Which environmental problems are of highest priority for our action?" we are asking a "Sophie's choice" question, because that question in essence also asks "Which environmental problems are of low priority for action?"

On the face of it, the comparative-risk-assessment question seems rational. The reasoning goes like this: Our society is facing innumerable environmental problems. Government agencies do not have the budget to address them all. We therefore must prioritize those that are most important and tackle those first.

Using this reasoning and comparative risk assessment, a city, state, or national government agency or a tribe assembles a group of people who look at a range of environmental problems that the group is facing. The group then ranks them in terms of some criteria. Some of the criteria might be irreversibility of effects, seriousness of human effects, degree to which the different problems are soluble, and how much money and other resources it would take to solve them.

The purpose of the exercise is to assign higher priority for action to certain environmental problems and lower priority for action to other problems.

For instance, in the mid 1980s the U.S. Environmental Protection Agency asked its senior staff to evaluate 31 environmental problems in terms of their relative environmental risks. The lowest four rankings were given to new toxic chemicals, environmental releases of genetically altered organisms, consumer product exposure, and, lowest of all, exposure of workers to chemicals (USEPA 1987b). Later the EPA asked a group of industry, agency, and academic "experts" (the Relative Risk Reduction Strategies Committee) to rank the importance of environmental problems that the EPA faced (USEPA 1990). That committee ranked groundwater pollution, oil spills, and radionuclides as "relatively low-risk problems" and habitat destruction, species extinction, ozone depletion, and global climate change as matters of high priority.

The EPA has become a major promoter of risk ranking—within the EPA, by states and tribes, and internationally. (See, e.g., Roberts 1990; Habicht 1994; USAID 1990.)

There are some useful aspects to some comparative-risk-assessment projects. One of the most useful is that sometimes such projects involve a diverse group of people actually thinking about and talking about the multitude of environmental problems that exist. Another useful feature of some comparative risk projects is that the participants come to realize that there are considerations other than death or disease that make an environmental problem important to people. The California Comparative Risk Project, for instance, included these social-welfare criteria, each of which considers more than strictly biological, chemical, or physical harm (CCRP 1994):

environmental and aesthetic well-being
economic well-being
physical well-being
peace of mind
future well-being
equity
community well-being.

Despite the potential that comparative risk assessment holds for broad thinking about environmental problems, I would contend that the basic assumptions and the processes of comparative-risk-assessment projects are highly problematic.

Questionable Assumptions of Comparative Risk Assessment

There are at least three linked assumptions behind comparative-risk-assessment projects: that the society in which the project is taking place "has" many environmental problems, that it doesn't have the resources to address all these problems, and that therefore it should prioritize which environmental problems to address.

Many people would disagree with all three of these assumptions. Such people would contend that our society causes (and has caused) many environmental problems, that it is capable of addressing and perhaps preventing all environmental problems, and that it should prioritize which processes will most effectively involve the entire society in addressing all environmental problems.

Let us look at the assumption that our society "has" many environmental problems (vs. that our society causes and has caused many environmental problems) and the assumption that our society doesn't have the resources to address all these problems (vs. that our society is capable of addressing and perhaps preventing all environmental problems).

Humans, both individually and collectively (e.g., when organized into corporations protected by numerous laws) cause environmental problems. If humans cause problems by particular behaviors, they can avoid causing them by not behaving in those ways.

If, for instance, major soil erosion is caused by clear cutting of forests in a particular region, we could undertake forestry in that region without clear cutting. If we are contributing greatly to global warming by burning fossil fuels, we could move away from burning so many fossil fuels by practicing energy conservation, producing fewer unnecessary products, and relying more on solar energy.

Many serious environmental problems are caused by corporations that are allowed by law to use water, air, soil, and the bodies of humans and wildlife as free depositories for their wastes. We could change laws in ways that would require corporations to severely reduce their use of toxic chemicals, to publish environmental audits and suggestions for the least harmful available alternatives, and to retain all toxic wastes in containers on their own property.

Avoiding causing environmental problems isn't easy, and our society may not have the will to address their causes and consequences. But if humans are causing environmental problems by certain behaviors, they are certainly capable of behaving in ways that do not cause those environmental problems or in ways that greatly minimize them. It is simply disingenuous to claim that environmental problems somehow magically exist and that society cannot address them all.

It is true that some types of environmental degradation and some losses are essentially irreparable. For instance, we humans have caused the extinction of many animal and plant species. We have produced nuclear wastes that cannot effectively be eliminated or stored and which will continue to cause environmental and human-health problems for thousands of years. The inevitable migration of the existing highly persistent, toxic, bioaccumulative PCBs out of our waste dumps and into the marine ecosystem are dooming certain marine mammals (e.g., whales and dolphins) to continued bioaccumulation of these toxic chemicals in their food chain and in their bodies. Their reproduction and development are thus threatened, even though they live far from the places where PCBs are produced and used. As one scientist noted in regard to marine studies of PCB contamination, "small cetaceans such as striped dolphins, melon-headed whales and Dall's porpoises were found to contain much higher levels of PCBs than terrestrial mammals and birds, in spite of living in the pristine oceans far from the land-based PCB pollution sources" (Tanabe 1988).

In such cases, the best we can do is try to ameliorate the consequences. We can't bring extinct species back, but we can make more adjustments in our activities so as to provide more "living room" for other species. We can't "clean up" all the harmful wastes we have produced, but we can commit major attention and funds to locating, storing, and containing nuclear wastes (USDOE 1995) and PCBs.

If all environmental problems can be addressed in some manner by changing the behaviors that are causing the problems, then ranking environmental problems for priority attention is less useful than determining which social arrangements and processes will lead most effectively to changes in the behavior of those who are causing the problems. Before discussing this alternative to comparative risk assessment, however, let us look at another set of problems with comparative risk assessment: its processes.

Predicaments of Comparative Risk Assessment

Trying to rank "environmental problems" leads to several obvious predicaments.

1. How does one define an "environmental problem" for the purpose of ranking?

There is no objective way to draw boundaries around a specific ecological or human-health problem for the purpose of establishing relative risk.

For example, how do we delimit the problem of "global warming"? Do we consider overpopulation as a part of global warming, or as a separate problem? As a national Committee on Science, Engineering, and Public Policy (1991) has noted, "even with rapid technological advances, slowing global population growth is a necessary component of a long-term effort to control worldwide emissions of greenhouse gases." Do we consider growth in per capita income a part of global warming? As the same committee notes, "reducing population growth alone, however, may not reduce emissions of greenhouse gases because it may also stimulate growth in per capita income" (ibid.). When the committee is referring to "growth in per capita income," it is essentially referring to growth in consumption of material products, which means growth in manufacture and use of products. Manufacture and use of products such as cars contributes to global warming through the burning of fossil fuels.

Can dioxin, one of the most toxic substances known, be considered a problem by itself? Do we rank it separately from pollution by other individual chlorinated toxics, such as chloroform, the herbicide 2,4-D, and chlorofluorocarbons?[1] Or do we combine these four and thousands of other chlorinated compounds and define dioxin as a subset of the chlorine problem? In reality, the only way to prevent continued production of dioxin is to phase out the use of chlorine as an industrial feedstock for pesticides, polyvinyl chloride (PVC) plastics, pulp and paper production, solvents, and other products. This is because producing chlorine and chlorinated compounds and burning chlorinated products all create dioxin.

Do we describe one problem as groundwater contamination and another as air pollution and a third as indoor air pollution, or do we instead lump all three as consequences of toxics use and then rank that as the problem?

It is interesting to note that the EPA Science Advisory Board did not rank the following as environmental problems:

• overpopulation
• overconsumption
• poverty
• legal protections for corporations as private persons and as toxic invaders of public and personal property such as air, water, and the bodies of humans and wildlife (Montague 1994b)[2]
• production of inessential products that result in environmental degradation.

Many would contend that all five of these are extremely high-risk environmental and public-health problems. There are political and economic reasons that problems such as these five were not included for consideration by the Science Advisory Board. But such politics simply illustrate that what is considered an environmental problem is *arbitrary*.

How environmental problems are defined and grouped during comparative-risk-assessment exercises is a major determinant of how they are ranked. Thus, ranking may be essentially an exercise in communal denial of fundamental environmental problems. If these problems continue to be ignored, attempts to address the lower-ranked problems may be limited to tinkering.

2. Who does the ranking?

Who is selected to rank whatever environmental problems have been identified?

Do those who profit from causing environmental problems sit in the group, or those who bear the damages? Who dominates?

Does the comparative-risk-assessment group include representatives of indigenous peoples who believe that the entire world is sacred? At a 1992 conference on EPA risk-ranking processes, one participant noted that uranium wastes should be of lower priority than certain other environmental problems because most of the wastes are out where nobody lives. If he were a Hopi or a Navajo from the southwestern United States, he probably would not have ranked uranium wastes that way. This is because uranium wastes have been dumped where the Hopi and the Navajo live, not where the speaker lives.

Does the group include factory workers from maquiladoras[3] (factories in a free-trade-zone in northern Mexico) who have become activists because they are being poisoned? Or is the group dominated by highly educated white scientists who have never heard of, let alone stepped inside, a maquiladora?

Is the group dominated by people who believe that economic and industrial growth are essential for a good human society? If so, the group will rank environmental problems quite differently than a group of people who are equally knowledgeable but who believe that endless economic and industrial growth is not possible without degradation of the environment and destruction of local communities.

3. Who is going to accept the ranking?

Even if comparative risk were a rational process, does anyone really think that citizens, acting through the democratic process, will accept a low ranking for an environmental situation that affects them directly? For instance, what self-respecting and aware worker would accept the EPA's ranking of "worker exposure to chemicals" as having the lowest risk of 31 environmental problems examined?

Likewise, if some group ranks a particular Superfund site as having a low priority for action, do you think the families that live next to that Superfund site will give it low priority? Should they, given repeated findings that people living near waste-disposal sites variously experience higher rates of birth defects, kidney disease, childhood leukemia, and so on (Montague 1994a)? I wouldn't pay attention for one minute to some risk assessors' ranking of a waste site next to my backyard as being of low priority for action relative to global warming. Would you?

Is it a responsible use of people's time to fight over which types of environmental degradation are worst and which can be ignored? Imagine if every time you felt you needed to see a doctor you had to go early in the morning to the doctor's office, along with everyone else who wanted to see the doctor that day. The doctor would ask each of you to describe your symptoms and would then decide whose symptoms were in the top third in terms of severity. The rest of you would simply have to leave.

Don't we have a sense that an ill person should be able to obtain some professional help? Do we think any less of the Earth than of our own bodies?

Every kind of environmental degradation matters to someone. Perhaps frogs do not matter to many people. Perhaps few people (see, e.g., Blaustein et al. 1994) regard loss of frog habitat, pollutant poisoning of frogs, and possible effects of the ozone hole on frogs as warranting any attention at all. However, others who care greatly about frogs, or who have a sense that what is done to the Earth is done to them, will seek to address the threats and losses that frogs are experiencing.

If a comparative-risk-assessment group ranks one particular environmental problem as being low priority, someone is sure to disagree, and that person will continue to fight for social attention to that environmental problem.

The fundamental flaw of comparative risk assessment, then, is not that it is arbitrary; it is the assumption that we should prioritize which environmental problems to address and, by extension, which ones not to address.

This leads us to the third pair of assumptions involving comparative risk assessment: the assumption that we should prioritize which environmental problems to address and the assumption that we should prioritize which processes will most effectively involve the entire society in addressing all environmental problems.

An Alternative to Comparative Risk Assessment

If we have a goal of preventing and solving environmental problems rather than choosing which ones to ignore, we will not ask which types of environmental degradation are worst. Instead, we will ask "How do we get industry, the government, and citizens to behave more carefully toward the environment?" This question focuses on conditions that encourage or require environmentally responsible behaviors.

It is noteworthy that comparative-risk-assessment processes rank environmental problems. It would be just as logical to rank which behaviors are causing the greatest environmental problems, or who is causing the greatest environmental problems, or which social arrangements allow or encourage people to cause environmental problems. By focusing on environmental problems rather than on problematic behaviors, problematic people, or problematic social arrangements, the comparative-risk-assessment group can pretend that the problems just "happened" and that no identifiable individuals or businesses caused them.

Ranking environmental problems rather than ranking problematic behaviors lets people avoid the reality that the problems were also caused in part by social arrangements, such as the legal framework for corporations that favors decisions that bring short-term profits over decisions that protect public-trust resources such as air and water (Montague 1994b).

As a result, the solutions to the environmental problems ranked high by comparative risk assessment will be primarily technological and "end-of-pipe," such as the following:

• better stack scrubbers, which will catch more of the toxic chemicals that are coming out of factory stacks

• better landfill liners, which will better contain toxic substances in the wastes dumped by factories and consumers

• better wetland construction techniques, which will reconstruct wetland habitat better when developers have eliminated other wetlands by building in them

• transporting wolves between states in airplanes to introduce genetic diversity to wolf populations that have been separated by tourist highways, logging roads, and other human developments.

Other solutions will be lifestyle adjustments, such as consumer recycling, which converts some consumer trash into other products. (Proposed lifestyle adjustments will likely not include lowered consumption, because that brings "economic growth" into question.)

If instead we rank the behaviors or social arrangements that cause the greatest environmental problems, we might include such things as these:

• laws and permits that allow businesses to release toxic chemicals into the public environment and that, furthermore, allow them to do so without telling the public what they are releasing

• the dependence of politicians on industry money to fund election campaigns

• laws that allow industry to use environmentally damaging technologies when much more environmentally protective technologies are available

• lack of involvement of public schools in teaching students how to become active citizens in this country (e.g., by writing laws, or solving environmental and political problems in their communities)

• the absence of environmental rights in the constitution

• low corporate taxes, which allow corporations to amass enormous wealth and political advantage over communities that will be or have been harmed by those corporations.

The disadvantage to industry and government of ranking which social arrangements will best achieve environmental health is that it suggests that social change is possible and desirable. Let us look at some rather simple social arrangements that might lead to prevention and reduction of an impressively broad range of environmental problems.

1. Require all businesses and government agencies to consistently and publicly explore, on paper, and in understandable language, their options for causing the least possible environmental damage.

All potentially environmentally degrading activities, public or private, should be subject to public scrutiny of alternatives. The public deserves to know that those who pollute, extract, consume, emit, incinerate, or abandon are aware of their technological options for minimizing disturbance of the environment. The public deserves to know what those technological options are, because they and the environment may suffer if a facility chooses a cheap but damaging technology over one that is more protective of workers and communities.

2. Provide citizens with easy access to all information relevant to effects on their environment and their health.

Informing citizens may be one of the most cost-effective routes by which to address all forms of environmental degradation. If state agencies are understaffed to monitor the environment, the general population is not.

For example, labeling products (e.g., listing all the ingredients in a pesticide formulation, or identifying milk from cows treated with bovine growth hormone) is not a high-cost process, and it certainly squares with free-market beliefs and with belief in the public's right to information. Given basic information about what is present in a product or where it came from, consumers can make choices on grounds other than messages from the seller.

How long would Florida tomato growers continue to use methyl bromide—a known ozone depleter—if they had to announce that fact on every box of their tomatoes? (All other products produced with other Class I ozone-depleting substances[4] have to be so labeled under the U.S. Clean Air Act. An exemption has been granted by the EPA for food products produced with methyl bromide (Clean Air Act 1993).)

Labeling of cement is another example. The owners of some cement kilns are paid by waste-disposal companies to burn toxic wastes (e.g., pesticides, tires, solvents) as fuel when heating limestone and clay or shale to make cement, while other operators pay for regular fuel to heat their kilns. Many industries that are disposing of hazardous wastes prefer sending their wastes to cement kilns rather than to hazardous-waste incinerators because it is cheaper to send the wastes to the cement kilns (which are not regulated as strictly as hazardous-waste incinerators). For instance, hazardous-waste incinerators have to place their incinerator ash in a licensed hazardous-waste dump, while cement kiln operators can leave their toxic-laden incinerator ash sitting out in the open air. According to one report, "90 percent of the liquid hazardous waste incinerated in this country and two thirds of the sludge and solid hazardous waste are burned in cement kilns" (USEPA 1993c). If their cement were labeled as having been produced with and as potentially containing dioxin and heavy metals, there would be market pressure on cement kilns to use regular fuel. Indeed, several communities (e.g., New Braunfels, Texas; Allegheny County, Pennsylvania) have passed resolutions to stop purchasing cement produced with hazardous waste.

Consumer labeling could be required for numerous types of products produced with technologies that are unnecessarily harsh on the environment.

What if the federal government, or state or local governments, were to produce user-friendly pamphlets for the public with such titles as "How you can find out what pesticides are being released on the farms near you and easily locate all human studies on effects associated with these pesticides," "How you can match toxic materials used by industries in your region with toxic substances found in your groundwater," and "How you can find out which birds of prey are in trouble in your area and the environmental pressures to which each of these birds are most vulnerable"? Any one of these pamphlets would make citizens much more likely to become angry at damage that is occurring and to make the case for change. Requiring that citizens be informed about environmental threats to them does not substitute for requiring more environmentally sound behavior by those who are creating the threats. However, when environmental information is readily available and complete, citizens and consumers in one location are often quite powerful at forcing change that has implications for other locations. Not incidentally, it is extremely difficult to even make the case for change when access to information is limited.

3. Provide for citizen-suit enforcement of all environmental laws.

No environmental law on Earth will be enforced adequately by any government if citizens do not have the potential to enforce that law. State agriculture departments, for instance, are generally less than enthusiastic about enforcing the federal pesticide law (Hassanein 1989). There are no citizen-enforcement provisions in that law.

Businesses and agencies are likely to break environmental laws and to cause environmental problems if they think there aren't many people looking.

Citizens must have legal and financial access to courts so as to safeguard the enforcement of environmental laws. The Administrative Procedures Act, for instance, provides citizens with the right to go to court to seek injunctions against actions by federal agencies that are "arbitrary and capricious," "not in accordance with law," or "without observance of procedure required by law" (5 U.S.C. § 706(2)). The Clean Water Act allows citizens to sue private companies and government agencies that violate Clean Water Act regulations.

The Equal Access to Justice Act requires the federal government to pay the fees of attorneys who bring and win cases against it on behalf of citizens.⁵ Thus, when citizens know that a law is clearly being broken by the federal government they can often find an attorney who will present their case without charging them, because attorneys who feel they are likely to win a case know they will be able to recover fees from the government. If the Equal Access to Justice Act did not exist, many citizen groups that do not have much money would not have actual access to courts even if they had legal access.

If a federal or a state agency doesn't have enough people to look into whether laws it administers are being broken, there are enough people in the country or in the state to do it. Impulses to protect one's locale, one's body, one's children, one's air, one's water, and one's non-human neighbors are deeply felt by many citizens.

I do not claim that every community has a large cadre of citizens who will take the responsibility for enforcing all environmental laws. A few effective lawsuits, however, can encourage change in the practices of many businesses and many agencies.

4. Encourage schools and universities to involve students in solving community and state environmental problems.

There is no larger source of unbounded energy, idealism, and creativity than students, and yet we have consistently failed to give our students practice at being active citizens. We have failed to help them learn how to look around, see what doesn't seem right to them, envision solutions to community problems, and pitch in to solve them.

What kind of citizen resources and citizenship literacy would we have available in our states for solving environmental problems if all our citizens had started practicing community and state environmental problem-solving in fourth grade?

Barbara Lewis, an elementary school teacher in Salt Lake City, helped her sixth-grade students in a low-income neighborhood as they investigated a waste site, got it listed as a Superfund site, and eventually wrote a bill (which became a state law) for generating funds to deal with waste sites. Ms. Lewis's *Kid's Guide to Social Action* (1991) is an amazing how-to book for telephoning, speaking, petitioning, fundraising, advertising, picketing, and initiating or changing local, state, and national laws. One section even describes how to amend the U.S. Constitution. *Civics for Democracy* (Isaac 1992), a similar book for older students, reviews the history of citizen movements in the United States and the techniques of participation.

Note that the four suggestions above involve little or no cost to governments. All four involve broadly distributed responsibilities, and all provide for public involvement in bringing social change for environmental responsibility. None of them involve ignoring any form of environmental degradation.

II

Alternatives Assessment as an Alternative to Risk Assessment

9

Alternatives Assessment: The Case of Bovine Growth Hormone and Rotational Grazing[1]

Alternatives assessment means looking at the pros and cons of a broad range of options.

My sister, a psychiatric social worker, has told me that one of the signs that a client might be suicidal is when the person is convinced there are only one or two options for her or his life, and both options are terrible.

Fortunately, most of us feel that we have some good options. We feel as if there is something we can do to relieve situations we find difficult. We assess our alternatives and choose the alternative that seems best to us.

Replacing risk assessment with alternatives assessment involves the same simple principle. Instead of allowing ourselves to be limited to one or two options that are terrible, we can insist on public consideration of a range of alternatives that seem good for different reasons. We can evaluate these alternatives and choose the one that seems best.

When evaluating alternatives, we do assess risks, but we assess the risks of a wide range of options. And we assess benefits of these same options.

Recombinant bovine growth hormone (rbGH) is a genetically engineered growth hormone that can be injected into cows to induce them to produce larger quantities of milk for a period of weeks.[2] A dairy cow's milk cycle lasts about 305 days, including a "rising phase" during which the cow secretes a large amount of natural growth hormone, a short peak phase during which milk production reaches its highest level, and a phase of declining milk production. Injections of rbGH are used to extend the peak phase and slow the decline. A dairy cow that had been producing 6 gallons of milk a day without rbGH might produce 6.6–7.5 gallons, an increase in milk yield of 10–25 percent.

Average "high yield" cows in the United States produce about 16,000 pounds of milk a year, while "superior genotypes" produce 26,700 pounds of milk per cow per year. Some in the dairy community think that cows bred to be extremely high milk producers are already at their physical limit, experiencing stress evidenced by increased disease and reproductive damage (Hansen 1990).

Use of rbGH is being hotly debated at the present time. Some people feel that it is harmful to cows, that it poses threats to humans who drink milk from rbGH-treated cows, that it exacerbates milk overproduction, and that it benefits large dairies at the expense of smaller ones. At present, sellers of milk produced using rbGH are not required to label the milk as such. This means that consumers generally cannot choose to buy milk that was produced with or without rbGH.

In addition, rbGH involves genetic engineering, a controversial technique by which genes from one living organism are inserted into another living organism.[3] The second living organism "accepts" the foreign gene into its genetic code and thus becomes an altered organism.

Because some scientists, many citizens, and many dairy farmers are questioning and/or opposing the use of rbGH, many questions are being asked about risk assessment, including these:

• How much more mastitis (inflammation of the udder) do cows experience when rbGH is administered? Some farmers and some scientists are finding that dairy cows injected with rbGH experience more mastitis. This question is of concern primarily because the cows experience discomfort and may be killed earlier if they contract mastitis, and because the farmer injects cows suffering from mastitis with pharmaceutical drugs, which may enter the cow's milk.

• How resistant is rbGH-associated mastitis to drugs? The development of drug-resistant organisms is of concern because some human and animal diseases are becoming difficult to treat with drugs.

• What are the amounts of drugs used to treat cows on farms using rbGH versus on farms not using rbGH? The major concerns here are whether consumers of rbGH-produced milk take in small amounts of pharmaceutical drugs, and whether disease organisms in those consumers may then develop resistance to those drugs.

• What methods will detect the presence of rbGH in milk? This is of concern because rbGH might cause adverse effects in humans who drink milk containing it.

• How much more insulin-like growth factor 1 (IGF-1) is present in milk from cows injected with rbGH than in milk from cows not injected with rbGH? Cow IGF-1 is a chemical "messenger" that is genetically identical to human IGF-1. IGF-1 produces ("mediates") much of the cellular response to rbGH in cows and to growth hormone in humans. There is some evidence that there are higher levels of IGF-1 in the milk of cows treated with rbGH. This is of potential concern if our bodies are trying to precisely control our natural amounts of IGF-1 and if we then add "extra" IGF-1 to our bodies by drinking cows' milk. There is much that we do not know about all the roles that hormones play in our bodies.

• Is -3N:IGF-1 present in the milk of rbGH-injected cows? If so, how much more of it is present in such milk than in the milk of cows not injected with rbGH? -3N:IGF-1, a slightly modified form of IGF-1, is normally present in a cow's colostrum—the yellow fluid, rich in protein and immune factors, that is secreted during the first few days of lactation. -3N:IGF-1 alters genetic processes by stimulating DNA synthesis. This question raises the concern that -3N:IGF-1, a growth-altering and gene-altering substance, might be introduced into the bodies of humans who drink milk from rbGH-injected cows.

Depending on the answers we get to these risk-assessment questions and the degree to which we are concerned about activities such as consuming extra human growth hormone in cow's milk we drink, we can make decisions on whether we will legalize or consume milk from rbGH-injected cows.

But we can also ask questions about alternatives to the risks of giving cows mastitis and feeding humans excess growth hormones. We could, for instance, compare reliance on rbGH with an alternative technology for increasing dairy farm profits: rotational grazing.

When grazing cows move from one pasture to another during a season, in essence they harvest most or all of their own food from perennial pasture crops. For most or all of the year, they are not confined in a barn or stockyard, to be fed crop plants that have been grown elsewhere and transported to them. (Cows injected with rbGH are, for the most part, raised within confined feeding systems.) In addition, by moving from pasture to pasture, the cows spread all or most of their own manure on the pastures during much of the year. In a confined feeding system, the manure becomes concentrated in one spot and must be spread mechanically by the dairy farmer. If there is not enough land on which to spread manure, disposing of it is difficult. Nitrate-rich manure can pollute rivers and underground water supplies.

Rotational grazing and rbGH, then, are both technological approaches to milk production. Though not technically mutually exclusive, in practice they are used separately. Because some farmers use rotational grazing[4] while many others use confined feeding and rbGH, there are documented and potential short- and long-term disadvantages and benefits to grazing cows rotationally and injecting cows with rbGH. (See box 9.1.)

When the two dairying options are considered side by side, a host of questions may arise:

• How much mastitis, and which types of it, are experienced by cows on rbGH farms and by those on rotational-grazing farms?

• How do symptoms of poor health differ between rbGH cows and rotational grazing cows? In addition to experiencing less mastitis, cows on rotational-grazing farms appear to suffer fewer other diseases. For instance, they have fewer hoof diseases because they don't stand on a cement floor.

• What is the amount and the nature of drug use on rbGH farms versus rotational-grazing farms? If the cows are ill less often or with different diseases on a rotational-grazing farm, they may be injected with fewer and/or different drugs.

• What is the fate of manure on an rbGH farm and on a rotational-grazing farm? For instance, does more manure enter water supplies from an rbGH farm?

• How much pesticide use is related to milk production on farms using rbGH versus on farms employing rotational grazing? rbGH is usually used in confined feeding operations in which cattle are fed indoors with feed grown elsewhere. The cows' food has usually been grown in single-crop (monoculture) systems, which often leads to intensive pesticide use.

• How does the condition of soils on farms growing feed for rbGH systems compare with that of soils grazed rotationally? Soil with perennial pasture plants and with a mixture of plants may be healthier than soil with annual crop plants or a single species of plant.

• Do farmers managing confined-feeding rbGH farms and farmers managing rotational-grazing farms differ in their work enjoyment? Several studies have indicated that farmers who manage rotational-grazing farms report higher work satisfaction or increased work satisfaction since abandoning management of a confined feeding system.

• What costs of regulation, research, monitoring, enforcement, and sur-

Box 9.1
An Alternatives Assessment: Recombinant Bovine Growth Hormone vs.
Rotational Grazing

Animal Health
Recombinant bovine growth hormone (rbGH) almost doubles the period during which a cow uses her own body tissues to make more milk. rbGH is associated with protracted infertility, mastitis (an inflammation of the mammary gland), and higher culling rates (killing of ill cows to remove them from the herd).
Rotational grazing improves herd health compared to confinement-feeding systems. There is higher reproductive performance, reduced lameness from leg or hoof problems, and fewer metabolic and digestive disorders.

Human Health and Consumer Response
rbGH requires greater use of antibiotics to treat extra mastitis, more monitoring programs to detect drugs in the milk.
rbGH releases another protein-hormone, IGF-1, in cows and increases it in milk; might cause problems in humans' upper gastrointestinal tract, especially in infants.
rbGH might cause immune and allergy reactions in people who drink the cow's milk.
Consumers want rbGH milk to be labeled.
Rotational grazing poses no additional risks to human health.

Economic Impacts
rbGH and rotational grazing each have the potential to increase profit and decrease feed costs, given specific conditions and circumstances.
rbGH aims to increase milk production faster than it increases feed and drug costs.
rbGH is increasingly competitive with rotational grazing under the following conditions: higher milk prices, low feed costs, low interest rates and capital costs,[1] low rbGH costs, high response rates of cows to rbGH.
Widespread adoption of rbGH would lead to more overproduction, which would increase government (taxpayer) purchases of surplus milk.
Rotational grazing emphasizes reduction of feed costs, but depends on pasture availability and fertility and the ability to maintain milk output.
Rotational grazing is increasingly competitive with rbGH under the following conditions: lower milk prices, high feed costs, high interest rates and capital costs, high rbGH costs, low response rates of cows to rbGH.

Economic and Social Viability of Rural Communities
rbGH will most likely benefit larger farms that can afford the associated costs and cause a decline in the number of mid-size dairies, particularly in the Midwest.
Rotational grazing encourages community self-reliance by minimizing purchased inputs.

Rotational grazing gives farmers greater flexibility than intensive confinement-feeding systems.[2]

Environmental Consequences
rbGH increases the risk of nitrate, herbicide, and insecticide contamination of water due to the increased cultivation of feed grain.
Rotational grazing results in 24 to 31 percent less soil erosion and 23 to 26 percent less fuel use in crop production. Pastures have double the organic matter content of land devoted to grain crops. This organic matter is a reservoir for carbon (carbon dioxide) and nitrogen (nitrates).

Social Context
Many argue that even if rbGH increases dairy profits and lowers consumer prices, its use reduces individuals' ability to participate in important personal and collective decisions.[3]

adapted from Liebhardt 1993

1. Rotational grazing is less capital intensive than confined feeding/rbGH systems. For instance, the farmer has to spend less money on storing feed in silos, hauling manure, or maintaining buildings for confined feeding, because the cows are out on the pastures, where their food is "stored," and they spread their own manure. This advantage is less of an economic advantage when costs of storage and hauling or interest rates on capital outlays are low.

2. For instance, some rotational graziers in Wisconsin are taking vacations during winter (rather unheard of for a dairy farmer) by producing milk primarily in the summer months when the pastures provide the best food, and their costs are least and profits greatest (Shirley 1993). As two observers of rotational graziers note: "The profound heresy in rotational grazing is that it has meant adopting a different way of thinking about production and profitability than that associated with the sociotechnical system of conventional dairying. . . . [For instance,] some grass farmers have remarked that adopting grazing made them subsequently question the need for [agrichemicals and petroleum] as they begin thinking about alternative means to profitability and utilizing their skills to create healthy diverse pastures" (Hassanein and Kloppenburg 1995).

3. For instance, rbGH dairy farmers are more dependent on corporate pricing of antibiotics and rbGH, more dependent on suppliers to research and "fix" the health problems of the herds, and more dependent on society's willingness to pay for surplus milk. Ultimately, rbGH farmers need to capitalize, expand, and capture more of the market in order to remain in the shrinking pool of dairy farms needed if cows overproduce milk.

plus milk purchases have been and will be borne by the public to support the use of rbGH versus to support of rotational grazing? If drug use and contamination of milk with drugs have to be monitored by the government, the monitoring will be paid for with tax money. If overproduction of milk is increased through the use of rbGH, the government will buy more surplus milk with tax money.

• How does consumers' approval of rotational grazing differ, if at all, from their approval of rbGH-based dairies? Several studies have shown consumers who are aware of issues surrounding rbGH and milk safety are concerned about the use of rbGH. Consumers overwhelmingly indicate they want rbGH milk labeled so they will be able to make their own decisions.

Some of the above questions, such as those about mastitis and other adverse symptoms in cows, are the types of questions that might well be asked in the course of a risk assessment of rbGH. However, some of the others (e.g., those about farmers' personal relationship to and enjoyment of farming,[5] about pesticide use, about the use of public money, about the long-term condition of agricultural soils, and about water quality) reflect larger concerns. These questions would not necessarily occur to us if we were doing only a risk assessment and looking only at confined feeding and rbGH use.

If we look at only the short- and long-term risks of rbGH (that is, if we do a risk assessment of rbGH), we get something like box 9.2. If, however, we compare the short- and long-term risks and benefits of rbGH and rotational grazing (that is, if we do an alternatives assessment of rbGH and rotational grazing), we get something like box 9.3.

Whenever an alternatives assessment is used to consider a broad range of reasonable options, two things are more likely to happen than when the risks of only one activity or a narrow range of options are examined:

• Because the differing benefits of the various alternatives remind us of divergent considerations, it is likely that we will ask a broader range of questions about the alternatives. Some of these questions will relate to social, democratic, economic, and political issues. In contrast, a risk assessment tends to focus narrowly on biological questions (e.g., toxicity, disease, mortality).

• The hazards of the more hazardous options are more likely to seem unacceptable, because those options seem more unnecessary when compared to reasonable alternatives that probably won't cause many hazards.

Box 9.2
A Risk Assessment for Recombinant Bovine Growth Hormone

Pros
None

Cons
Animal health
More mastitis
Higher culling rates

Human health
More use of antibiotics in cows
More monitoring programs to detect drugs in milk
More IGF-1 (a human hormone) in milk

Economics
More overproduction of milk
Increased government purchases of surplus milk

source: Mary O'Brien

If you wanted to get approval to undertake a particular hazardous activity, would you want people asking big questions about the activity? Would you want people to think that the hazards or the potential risks were unnecessary? Alternatives assessment threatens the status quo. Alternatives assessment can make social change seem both desirable and possible.

Consider some of the risk-assessment examples that were presented in chapter 3:

• What if the benefits and risks of incineration (e.g., of death from inhalation of particulates) were compared to the risks and benefits of reducing toxics use (which would result in less waste to be disposed of)?

• What if the benefits and risks of chlorine-based pulp production (e.g., killing bald eagles with dioxin) were compared to bleaching pulp with oxygen and hydrogen peroxide (which would produce no dioxin), or to making paper without wood?

• What if the benefits and risks of using dacthal to grow onions in eastern Oregon (e.g., contaminating drinking water aquifers) were compared to those of growing crops that do not depend heavily on herbicides or other pesticides?

• What if the benefits and risks of nuclear power (e.g., cancer in workers) were compared to the risks and benefits of energy conservation, wind power, and solar power?

Box 9.3
An Alternatives Assessment for Milk Production

Pros of bovine growth hormone (rBGh)
Economics
Increases milk production
Increases profits
Decreases feed costs
Benefits large dairies

Pros of rotational grazing
Economics
Increases milk production
Increases profits
Decreases feed costs
Benefits family farms

Animal health
Improves herd health (reproduction; reduced leg or hoof problems; fewer metabolic, digestive disorders)

Community/personal
Self-reliance: Minimizes purchased inputs
More farmer flexibility than with confinement feeding

Environmental
24–31% less soil erosion
23–26% less fuel use in crop production
Double organic matter in pasture crop land

Cons of bovine growth hormone (rBGh)
Animal health
More mastitis
Higher culling rates

Human health
More use of antibiotics in cows
More monitoring programs to detect drugs in milk
More IGF-1 (a human hormone) in milk

Economics
More overproduction of milk
Increases government purchases of surplus milk
Benefits large dairies at expense of family farms

Environment
Increases pesticide use/pollution to cultivate grain

Cons of rotational grazing
None

• What if the benefits and risks of tour-bus and recreational-vehicle traffic on the narrow western rim of Hells Canyon (including disruption of elk migration routes) were compared to the risks and benefits of encouraging walking, bicycling, and infrequent public van runs on the rim during summer months?

In each of the above cases, those who wanted to undertake a hazardous activity would strive to prevent the assessment of reasonable alternatives.

Those who work to have our society live within the limits of nature on Earth are too easily defeated by risk assessment. As Langdon Winner observes in *The Whale and the Reactor* (1986), "the risk debate is one that certain kinds of social interests can expect to lose by the very act of entering." Here Winner is talking about the difficulty of citizens and consumers entering into technical risk-assessment arguments about the presence and the risks of -3N:IGF-1 in cows' milk, the statistical likelihood of mastitis in cows, and the likelihood that particular disease organisms will develop resistance if particular drugs are used on certain dairy farms. Corporations that have enormous profits riding on the regulations that result from risk assessments will pour money and expertise into the assessments. Consumer groups that begin to question the wisdom of using rbGH for any of numerous reasons (including that they won't be able to know whether the milk they buy was produced using rbGH) will generally not be able to match the money or the expertise going into the defense of rbGH, which is being carried out almost exclusively on risk-assessment grounds.

Consumers, however, are much more likely to participate if the debate considers farmers' satisfaction, the common sense of rotational dairy farms, and the possibility of avoiding drinking pharmaceutical drugs.

Citizens are more likely to challenge special interests that intend to threaten the environment and the public health for no justifiable reason when good alternatives are laid out in front of them.

10
Alternatives Assessment vs. Cost-Benefit Analysis: There Is More to Life Than Money

Cost-benefit analysis considers only pros and cons that can supposedly be reduced to money. Alternatives assessment includes additional pros and cons that matter to people and the environment.

Think of any financial decision you have made—for instance, a job decision, a time you bought a book you had heard was good, a decision you made about whether to have a child, or a decision as to how much money to give to a certain organization.

If the only considerations in your decision were dollar considerations, you made your decision on the basis of cost-benefit analysis. If any other considerations entered into your decision, such as a sense of goodness or longing, a feeling for a place, a sense of fairness to your partner, or a belief in democracy, then you did not act solely on the basis of a cost-benefit analysis. Non-financial considerations played a role.

Cost-benefit analyses are used in social decision making in our society when the decision making is based on, or at least considers, what money might be gained and lost as a consequence of particular actions or regulations. Someone might, for instance, do a cost-benefit analysis of limiting the amount of benzene (a cancer-causing chemical) to which workers may be exposed in their workplace. The financial benefits to the employer of a moderately high level of workers' exposure to benzene might include savings from not buying expensive containment equipment. The costs to an employer of a moderately high level of workers' exposure to benzene might involve paying a higher premium for workers' compensation insurance, although this higher premium rate would be spread across many industries.

On the other hand, a worker might not experience any direct financial benefit from a moderately high benzene exposure limit but might experience direct financial costs as a result of contracting cancer.

If only one option (e.g., a certain exposure limit for benzene) is analyzed in a cost-benefit analysis, that is not a form of alternatives assessment. At other times, however, some of the comparative financial costs and benefits of more than one regulation, policy, or course of action are analyzed to see which alternative yields the largest net financial benefits. In this case, a single "dollar net benefit" (or "dollar net cost") is used to summarize each option. Such cost-benefit analysis is a form of alternatives assessment, because it considers the financial pros and cons of alternatives. The range of alternatives compared may be narrow or broad; however, the only benefits considered are benefits described as money, and the only costs considered are costs described as money.

A cost-benefit alternatives assessment of managing a dairy that uses rbGH and confined feeding versus managing a rotational-grazing dairy, for instance, would consider such financial elements as the costs of feed, its transportation, and its storage; the costs of buildings; the costs of antibiotics; the culling of cows; interest rates; the price of milk; and public subsidies. It would not consider job satisfaction unless a dollar figure were assigned somehow to that on the basis of some assumption. It would not consider a sense of independence from drug companies and genetic engineering experienced by the operator of a rotational-grazing dairy. It would not consider a community's perception of its independence of non-local inputs into dairy operations (e.g., growth hormones, antibiotics, non-local feed).

The Inability of Monetary Cost-Benefit Analysis to Account for Worth

The commonest deficiency of cost-benefit analysis is that not all benefits and costs of actions can be reduced to money. After all, one dictionary definition of "cost" is "the price paid to acquire, produce, accomplish, or maintain anything," and another is "a sacrifice, loss, or penalty" (Costello 1991). Clearly a price can be something other than money; it could be your life or your personal sense of integrity.

What is the dollar value to you, for instance, of your not having to watch your child die of brain cancer while you wonder if the cancer is due to toxic chemicals you brought home on your work clothes? Note that a World Bank memo (Summers 1991) urged the encouragement of more migration of "dirty industries" to Least Developed Countries in part because the economic earnings of the people there are less: "The measurement of the costs of health impairing pollution depends on the foregone earnings from increased morbidity [illness] and mortality [death]." What is the dollar benefit to the seventh generation from now of hearing wolves howl? What is the dollar benefit of having participatory democracy? What has been the dollar value to you of hearing frogs call in the evening?

The failure of cost-benefit analysis to include non-monetary considerations is the main basis for rejecting monetary cost-benefit analysis as a sufficient basis for any environmental, public, or private decision making.

Using monetary cost-benefit analysis for environment protection ignores several realities[1]:

- Many public goods are not available in the marketplace. Public goods (as opposed to private property) are goods that are available for enjoyment by all. You, individually, may be able to buy a diamond ring or a loaf of bread, but you cannot buy an independent justice system or a wilderness system, or the international behaviors that will produce clean rivers or an intact ozone layer.

- Many public goods are prevention of loss or harm, rather than tangible goods that can be consumed or added to our stock of wealth. While a person may buy a gold ring, that person has a harder time buying the prevention of ecological harm to watersheds through mining laws that prevent heap-leach gold mining.[2] Nearly every major environmental law (i.e., an example of a public good) is intended primarily to avoid or minimize losses or protect resources from harm.

- Many elements of life have no price tag. What is the dollar value of the existence of the spotted owl or lynx? What is the dollar value to a woman of not getting breast cancer? What is the dollar value of silence? Recently I climbed up Skinner's Butte, a city park in my town, Eugene, Oregon. Once on top, I rested under a large, old oak. I realized that the loud sounds far below of traffic from the Beltline and other streets in filled the entire air column above the town. I knew I would probably never be able to sit on top of Skinner's Butte and find relief from traffic noise. What is the dollar value of freedom from constant motorized noise? What is the dollar value of such deep darkness that you can see the Milky Way reach down to both horizons?

• The marketplace price of many elements of life does not reflect their value. Diamonds may cost a great deal to buy and water very little, but which is of more value to you if you haven't had either for six days?

• Many elements of life can be irreversibly stolen from future generations.

Aldo Leopold (1941) once wrote that he "wouldn't want to be young again without wild country to be young in." In 2050, will any children have the chance to feel the power of a wild steelhead in a wild river, or to watch a mule deer bound along a pine-studded hillside?

In the "marketplace," goods can be replaced or substituted. Who, however, will be able to bring back clean land and clean groundwater on the Hanford Nuclear Reservation in central Washington at any time during the next few hundred years? Who will be able to buy a cooler climate two generations from now? Males in industrial countries have approximately half the viable sperm of men 50 years ago (Carlsen et al. 1992). What is the price of sperm to a developing male embryo?

Decision making on the basis of monetary cost-benefit analysis is inappropriate for other basic reasons, too:

• Those who reap the monetary benefits are not necessarily those who pay the monetary costs. The carcinogenic pesticide acetochlor, for instance, may bring profits to DuPont, but DuPont will probably not pay for the breast mastectomies that acetochlor might necessitate (Paulsen 1994). While aluminum companies profited from the cheap electricity generated by dams on the Columbia River, Native American tribes in the Columbia River Basin lost salmon as a subsistence food (White 1995).

• "Cost," like "risk," is defined politically by those who have the power to define the numbers. The price paid for hydroelectric power by an aluminum corporation, for example, is not the price that would be set by Native Americans or others trying to save salmon and their cultures from extinction in the Pacific Northwest. The cost charged by Congress to a ranching family for grazing a couple of hundred cow-calf pairs on thousands of acres of dry public grasslands is not the cost that would be set by botanists, fish and wildlife biologists, wetlands specialists, and many citizens who see critical wildlife habitat being compromised or eliminated.

• The cost of a newly developed, environmentally protective technology may not be the cost of that technology a few years from now. When a technology becomes more widely available or when industry learns how to improve on the technology, the price can come down considerably. Therefore, a decision not to require a protective technology because it "costs" too much may be fundamentally flawed in the long run. This problem is magnified, of course, by the fact that the protective technology may bring long-term (even inter-generational) benefits that have no price tag.

Alternatives Assessment vs. Cost-Benefit Analysis

Nicholas Ashford and Charles Caldart (1991), who study technology, innovation, public policy, and law, describe three main advantages that supposedly characterize cost-benefit analysis as a decision-making tool. The first two advantages are shared with alternatives assessment; the third is not an appropriate "advantage" to pursue:

First, cost-benefit analysis clarifies choices among alternatives by evaluating consequences in a systematic manner. Second, it professes to foster an open and fair policy-making process by making explicit the estimates of costs and benefits and the assumptions on which those estimates are based. Third, by expressing all of the gains and losses in monetary terms, . . . cost-benefit analysis permits the total impact of a policy to be summarized using a common metric and represented by a single dollar amount.

Let us look at these three "advantages" one by one.

Alternatives assessment, like cost-benefit analysis, "clarifies choices among alternatives by evaluating consequences in a systematic manner" (Ashford and Caldart 1991). The consequences (i.e., the pros and cons, or the costs and benefits) of alternatives assessment, however, can include issues of democracy, aesthetics, spiritual values, ethnic values, uncertainty, sense of community, and personal feeling as well as monetary consequences. What constitutes "systematic evaluation" of these consequences, of course, is a judgment call. What may be "systematically evaluated" as crucially important to some may be "systematically evaluated" as marginally important to others. For instance, to not pass on a life integrated with salmon means a devastating loss of culture and identity to members of certain Native American tribes, to others who fish for a living, and to others whose lives are intertwined with a river rhythm. However, to some businesspeople and to some grocery shoppers it may mean only an insignificant change of fish on the dinner plate. Considering and comparing monetary consequences with aesthetic or cultural consequences may not appear as "systematic" as comparing dollars to dollars, but it is more realistic than assuming that all significant consequences can be systematically translated into dollars. To speak of making decisions "rationally" on the basis of money is to deny the reality that much that matters has no price tag.

Furthermore, like cost-benefit analysis, alternatives assessment "fosters an open and fair policy-making process by making explicit the estimates of costs and benefits of the options being considered and the assumptions on which those estimates are based" (Ashford and Caldart 1991). The costs and benefits considered in alternatives assessment, however, are more comprehensive than those considered in cost-benefit analysis. In an alternatives assessment, costs and benefits are both monetary and non-monetary. An alternatives assessment to which the public contributes will be much more candid than a cost-benefit analysis, because many kinds of relevant pros and cons will be displayed for public consideration.

The third "advantage" of cost-benefit analysis, i.e., the representation of the "total impact" of a policy or action by a "common metric" of dollars (Ashford and Caldart 1991), is not shared with alternatives assessment, because social and environmental realities cannot be reduced to a common metric such as dollars. Again, the tidy appearance of a common metric can be created only by denying social, political, spiritual, and environmental realities. Life on a Midwestern tall-grass prairie on a summer day cannot be represented by a single dollar value. That the "total impact" of any policy or action can be represented by dollars is illusory.

How, then, can we decide among the alternatives if we have to consider numerous groups and numerous factors, some of them financial, some cultural, some spiritual? The same way each of us makes decisions each day. Our decisions are sometimes based on sheer survival instinct, sometimes on overriding economic reasons, and sometimes on personal principles. Occasionally we choose one alternative because another alternative simply doesn't seem to be the right one. Sometimes we choose an alternative extremely systematically, on the basis of explicit factors, after careful consideration of numerous options and after consultation with others.

Social entities such as communities, school boards, legislatures, agencies, and corporations likewise make decisions on the basis of numerous factors. Political pressures will surely play roles, and one role may be to dress up other decision-making processes so that they appear "objective."

However, the choice of what dollar amount to put on the value of a worker's not getting cancer, for instance, is hardly non-political. To see the political nature of a cost-benefit analysis, simply ask the head of a corporation what dollar value should be placed on a worker's not getting cancer, then ask the worker.

Although both "objective" cost-benefit analysis and messier alternatives assessment are political, cost-benefit analysis covers up the political nature of analysis with monetary numbers. Alternatives assessment forces the decision maker to assume responsibility for choosing among various explicit political and value tradeoffs. As Ashford and Caldart (1991) note, "tradeoff" analysis (equivalent to alternatives assessment) forces decision makers "to make explicit their value judgments and tradeoffs, thereby preventing them from abdicating responsibility for their decisions." Here Ashford and Caldart are speaking to the heart of good public decision making: Governmental, business, and corporate decision makers must take responsibility for their decisions that affect the public and the public's environment. They must be prevented from hiding behind "dollar" numbers, just as they must be prevented from hiding behind "risk" numbers.

11

We Already Know How to Do Alternatives Assessment

Alternatives assessment is not a new process. Industrialized societies have occasionally established and implemented processes for doing it, and have learned to do it in a number of settings. The challenge before us is to greatly expand its use.

Alternatives assessment can be installed as a sensible three-step public process for making decisions about all behaviors that affect the environment. The steps are these:

• Consider a range of reasonable alternatives.
• Discuss the potential environmental, public-health, and social benefits of each alternative.
• Discuss the potential adverse environmental, public-health, and social impacts of each alternative.

The following are examples of social (as opposed to purely personal) alternatives assessment.

The National Environmental Policy Act

Although the National Environmental Policy Act (NEPA), passed in 1969, is quite brief, a set of regulations has been prepared for federal agencies to follow in order to comply with it. These regulations govern the preparation of environmental impact statements (EISs) by federal agencies (Council on Environmental Quality 1992b). The NEPA regulations were prepared in 1978, and only one regulation (regulation 1502.22, which required agencies to consider the worst impacts that could reasonably happen) has been slightly altered since that time. The regulations (with the exception of the new section 1502.22) are a model of clear language and good, public interest and environmental planning, and they require and respect public participation.[1]

NEPA regulation 1502.13, titled "Alternatives Including the Proposed Action," prescribes a process for keeping alternative approaches to the environment in mind during democratic decision making:

This section is the heart of the environmental impact statement. . . . It should present the environmental impacts of the proposal and the alternatives in comparative form, thus sharply defining the issues and providing a clear basis for choice among options by the decision maker and the public. In this section agencies shall:

(a) Rigorously explore and objectively evaluate all reasonable alternatives, and for alternatives which were eliminated from detailed study, briefly discuss the reasons for their having been eliminated.

(b) Devote substantial treatment to each alternative considered in detail including the proposed actions so that reviewers may evaluate their comparative merits.

(c) Include reasonable alternatives not within the jurisdiction of the lead agency.

(d) Include the alternative of no action.

Under NEPA regulations, federal agencies are not required to select the most environmentally sound alternative they describe, but they are required to discuss such alternatives publicly. Most simply, the NEPA regulations require federal agencies to "look before leaping," as Nicholas Yost, a government lawyer who oversaw their preparation, once noted (personal communication).

The NEPA requirement to consider "all reasonable alternatives" may sound like common sense. However, it is heartily resisted by those who wish to continue business as usual. Various reasonable alternatives propose changes in technologies, reductions in resource extraction (e.g., logging or mining), or even restrictions on the numbers of people who may undertake a given activity in an area.

The specter of fundamental change can be threatening to those who profit from particular ecosystems or natural resources, such as water, air, or oil. Certain reasonable alternatives can threaten entrenched social and financial arrangements.

If a group of citizens feel that a federal agency has failed to abide by NEPA regulations in the preparation of its environmental analyses, those citizens may first appeal to the agency itself to change its analysis. If that fails, they may sue the agency in a federal court.[2]

Unfortunately, courts have tended to be more concerned with agencies' failures to reveal environmental impacts of their proposed projects in EISs than with agencies' failures to adequately consider and present reasonable alternatives in their EISs. This is unfortunate, because a narrow range of alternatives will merely compare minor variations of an agency's proposed project.

Example: Permits for Grazing on Federal Lands

The multiple environmental impacts of livestock grazing on native, public grasslands in the western United States are receiving more and more public and scientific scrutiny. Among the concerns are degradation of streams, loss of nesting cover for ground-nesting birds, damage to rare plants, and denuding of soil.

In 1994, Tania Wolf of the Ochoco Resource and Recreation Association successfully appealed an Environmental Assessment[3] for a permit to graze cattle on the Ochoco National Forest of Oregon. Wolf contended that the Forest Service had not included in the Environmental Assessment the alternative of allowing no grazing in that area of the forest. She wanted the Forest Service to compare how much more quickly the land already damaged by cattle can recover if no cattle are grazing there than when cattle are grazing there.

In response to Wolf's appeal, the Forest Service's Regional Appeal Deciding Officer wrote the following (Ferraro 1994): "I find your issue regarding the need for a no grazing alternative has merit. NEPA requires that the Responsible Official choose among a reasonable range of alternatives and determine the significance of the chosen alternative." This might seem a small victory for the environment to those who are not familiar with cattle and sheep grazing on public lands in the western United States, but the mere discussion of the no-grazing alternative puts in high relief the reality that private grazing on public lands, with its multitude of environmental impacts, is not a foregone conclusion. It is not a legal right of private ranchers; it is a societal choice regarding the appropriate disposition of public forests, grasslands, deserts, rivers, and springs.

Example: Herbicide Spraying of National Forests

After allowing a private timber company to clearcut some acres in a national forest, the National Forest Service often sprays the clearcut area with herbicides to kill all vegetation that might compete with newly planted trees for sunlight or water. And the Forest Service sometimes sprays again, after several years, to "knock back" unwanted vegetation in the tree plantation.

In the late 1970s and the early 1980s, citizen organizations won several NEPA-based lawsuits because a Ranger District, a National Forest, and Region 6 of the Forest Service, all in the Pacific Northwest, had failed to adequately discuss the environmental and health impacts that such herbicide spraying could cause. A federal court finally enjoined Region 6 (Oregon and Washington) of the Forest Service from using any herbicides pending a rewrite of that Region's environmental impact statement for vegetation management (*NCAP v. Lyng*, 673 F. Supp. 1019 (D. Or 1987), *aff'd*, 844 F.2d 588 (9th Cir. 1988); see O'Brien 1990b).

During the injunction period, the Forest Service had to rely on non-chemical methods to clear unwanted vegetation, and eventually it wrote an EIS that adequately discussed alternatives to the use of herbicides on the forest lands it managed. Eventually, the Forest Service adopted as its preferred alternative one that placed explicit priority on preventing unwanted vegetation and on non-chemical treatment of vegetation.

As Thomas Turpin, then Forest Silviculturist (timber-production manager) for the Siuslaw National Forest, noted in a speech to foresters in British Columbia, the entire NEPA process resulted in changed assessments of the viability of non-chemical vegetation management alternatives (Turpin 1989):

When public pressure over herbicide use began to increase in the mid 1970s, we [the Forest Service] took the attitude of "We're right," "We're the professionals," [and] "We know what's best." We took the stance of discounting or discrediting public comments against the use of herbicides, and at times we also discredited the individuals involved. . . .

After the first lawsuit the Forest Service kept on using herbicides with little change in either the program or attitude. We continued to deny that any non-chemical methods would work. We felt that these other methods were too expensive and ineffective and that unacceptable tree mortality would result. . . .

In the Siuslaw National Forest, where the use of herbicides was the standard method used in the control of unwanted vegetation, the reality of not having [herbicides] began to sink in. This realization forced us to find successful non-chemical methods to treat vegetation problems.

Although some non-herbicide trials were established prior to 1984, the injunction pushed our search for alternative methods to the forefront. During the injunction period we developed, and are refining, the following techniques. . . .

The techniques included different seedling specifications, defining tolerable limits of vegetation competition, reforesting with no burn, and cutting salmonberry during specified seasonal "windows of vulnerability" so that it will not sprout again.

Some people question the power of lawsuits to "solve anything." However, before litigating, the Oregon citizen organizations had unsuccessfully appealed numerous herbicide spraying projects at the administrative level, and the Forest Service had routinely denied the appeals (SOCATS 1983). Had citizens not sued the Forest Service in several courts for its failure under NEPA to candidly discuss herbicide dangers, and had the Forest Service not been required by NEPA regulations to consider alternatives to herbicide spraying, it is not likely that the Forest Service ever would have shut down its spray nozzles long enough to consider the merits of reasonable alternatives.

The Endangered Species Act

Section 7 of the Endangered Species Act [16 U.S.C (1982)] requires all federal agencies to ensure that they do not jeopardize the existence of an endangered species or destroy habitat that it needs to survive. If a proposed action by a federal agency (e.g., building a road near a stream bearing endangered bull trout) is found to place a species' survival in jeopardy, the Secretary of Interior must issue a biological opinion suggesting "reasonable and prudent alternatives" that "can be taken by the agency or applicant." A federal agency, an applicant, or a state agency may then apply for an exemption to this biological opinion, but consideration of alternatives is again foremost. An exemption committee may grant an exemption only if at least five of its seven members find that there are no reasonable alternatives to the action that threatens the species, that the benefits of the proposed action outweigh the benefits of alternative courses of action that would conserve the species, and that the proposed action is of regional or national significance (Houck 1989).

The requirement to take a hard look at available alternatives has meant that exemptions to biological opinions finding species in jeopardy have been rare (Houck 1989). Humans can generally find alternatives to purposely causing a species to go extinct.

Consumer Reports: Alternatives Assessments of Consumer Products

Millions of U.S. consumers rely on the assessments of alternative consumer products published in the magazine *Consumer Reports*, published by the Consumers Union. An article on vacuum cleaners, for instance, will discuss both drawbacks (risks) and advantages (benefits) of a range of available vacuum cleaners.[4]

Sometimes *Consumer Reports* assesses alternative interpretations of scientific information. An article titled "Electromagnetic Fields" (*Consumer Reports* 1994b) asked: "Should you worry about them? Scientists still don't know for sure. We tell you what studies have shown—and give prudent advice."

The key to the ability of the Consumers Union to produce alternatives assessments is its nonprofit status and its policy of independence: "We buy all the products we test on the open market. We are not beholden to any commercial interest, and we don't take outside advertisements. Our income is derived from the sale of *Consumer Reports* and our other publications and products, and from nonrestrictive, noncommercial contributions, grants, and fees." ("About Consumers Union," Contents page of *Consumer Reports*, May 1994).

The public ends up bearing the consequences of public decisions regarding hazardous activities by "private" business and government agencies. In this sense, members of the public are involuntary consumers of those decisions. This is why our society must take a *"Consumer Reports* approach" to decisions that will affect the environment. The public needs to be able to assess the advantages and the disadvantages of a wide range of alternative policies and technologies, and to participate in decisions about those alternatives.

The Public-Health Model: Alternatives for Prevention of Disease

Public health, according to John Last, editor of the encyclopedic *Maxcy-Rosenau Public Health and Preventive Medicine* (1980), can be defined as "the combination of sciences, skills, and beliefs that are directed to the maintenance and improvement of the health of all the people" and as "concerned with the well-being of mankind, and of course with the well-being of individual members of society." "Sometimes, however," Last notes, "situations arise in which community interests must override individual interests."

Last distinguishes primary, secondary, and tertiary prevention of physical, mental, and emotional disease and injury. Tertiary prevention means minimizing the effects of disease and disability. Secondary prevention means early detection and intervention of disease or injury. Primary prevention, however, means preventing the occurrence of disease or injury, whether viral, parasitic, or bacterial infections; food poisoning; diseases transmitted to humans by insects or mammals; occupational disease; or exposure to toxic chemicals, heavy metals, and radiation.

Primary prevention, then, necessarily focuses on assessing the health benefits of particular activities, not solely on reacting to disease or injury (i.e., assessing the damages or "risks" of particular activities or conditions).

Primary-prevention measures that have been taken within the U.S. public health system include immunizing children against polio, eradicating smallpox, prohibiting leaded gasoline, and banning the production of PCBs. Secondary-prevention measures include various kinds of screening and early detection, such as Pap smears to test for precursors of cervical cancer, tests for levels of lead in the blood of children, and periodic health checkups. Tertiary prevention includes all measures to help people with diseases, such as treatment of asthmatic people with pharmaceutical drugs.

It is interesting that Last notes that beliefs combine with sciences and skills to produce the practice of public health. One of these beliefs is that society can consciously and willfully prevent health damage and promote health. We have alternatives, then, beyond merely monitoring for environmental and public-health damage and beyond merely "mitigating"[5] damage to the environment and to public health. We have the alternative of *preventing* such damage in the first place.

The public-health model provides a guide to the range of alternatives that should be considered in an alternatives assessment.

Technology Options Analysis

Presumably drawing from the public-health model of primary prevention, Nicholas Ashford of the Massachusetts Institute of Technology has proposed moving industrial facilities from "secondary prevention and mitigation" activities to "primary prevention" activities in order to prevent chemical accidents (Ashford 1993a). According to Ashford (1993b):

Primary prevention relies on the development and deployment of *inherently* safer technologies that prevent the *possibility* of an accident. Secondary prevention reduces the *probability* of an accident. Mitigation and emergency responses seek to reduce the *seriousness* of injuries resulting from accidents.

In many cases, alternative production processes exist which completely, or almost completely, eliminate the use of highly toxic, volatile, or flammable chemicals. Normal accidents arising in these systems result in significantly less harmful chemical reactions or releases. Replacement of existing production systems by such benign chemical processes, as well as non-chemical approaches, are examples of primary accident prevention.

A major omission of the [1994 Environmental Protection Agency] proposed rule [on risk-management programs for chemical accidental release prevention] is that it does not require facilities to evaluate *alternative* production technologies and explain why they are not adopting them. . . . Secondary prevention perpetuates a focus by the firm on risk assessment rather than [shifting] attention to technological solutions.

Ashford cites several examples of primary prevention:

• Monsanto has halved its total volume of highly toxic gases in storage by shifting to just-in-time deliveries of raw materials.

• Dow reduced its 100,000-pound inventory of the extremely dangerous nerve gas phosgene by 95 percent by having its units run continuously off a feed unit rather than discontinuously out of storage.

• PPG recently developed carbonyldiimidazole, a "benign phosgene substitute" that can be used in the synthesis of some of the company's pharmaceutical products.

Ashford suggests that any firm manufacturing or using hazardous chemical[6] should be required to prepare an annual Technology Options Analysis (TOA). "The scope of the TOA," Ashford notes (1993b),

will depend on the technical options that are realistically open to the firm in terms of scale, performance, and cost. In general, for an existing plant, the options, by necessity, will mostly involve secondary accident prevention and mitigation technologies. . . . For some processes used in existing plants primary prevention options might also be considered, especially those involving materials substitution or minor process changes. While these changes might be considered "minor" from the perspective of disrupting production, they could yield significant safety benefits.

For new plant construction or for major modifications to an existing plant, the options will involve primary and secondary prevention (as well as mitigation). Where both continued use of an existing plant and major modifications are possible, technological options will range from primary prevention to mitigation.

The TOAs would be submitted to regulatory offices that would verify their adequacy. "However," writes Ashford, "most information in the Plan may also [be] required to be made available to the public," with the exception of trade secrets.

Ashford describes the functions TOAs would serve:

First and foremost, TOAs are a valuable source of information to the firm—which might not otherwise be cognizant of its technological options—and thus can help the firm to make sound technological choices. In addition, regulators, both in EPA and [Occupational Safety and Health Administration], may use the TOAs to develop primary accident prevention strategies and policies. . . . [The] process of conducting the TOA and reporting the results of the TOA would ensure not only that the firm recognizes the hazards to which it is exposing its workers and the surrounding community . . . but also that it recognizes superior technological options that are available.

Ashford describes other policies that should be in place to provide incentives to enact primary prevention. The federal government, for instance, might require firms to carry insurance adequate to pay for the damages their chemical accidents are judged to have caused. Firms emphasizing primary prevention would pay smaller premiums for their insurance coverage than those relying on secondary and tertiary prevention.

New York City's Community Right-To-Know regulations require facilities that store, handle, use, or process hazardous substances to employ a form of technology options analysis (New York City Department of Environmental Protection 1994). Section 41-11 (b)(1) specifies that every such facility must file a risk-management plan that considers "the use of alternative substances and equipment to eliminate or reduce the use of EHS's [extremely hazardous substances included on the city's list of highly toxic, flammable, or explosive chemicals] or regulated toxic substances." Thus, New York City has taken the step of requiring plants to consider alternatives that would better protect workers and their surrounding communities. Presumably any community could require this of facilities operating within its boundaries.

The Massachusetts Toxics Use Reduction Act[7]

In 1989, the Massachusetts legislature unanimously passed the Toxics Use Reduction (TUR) Act, which promotes reducing the use of toxics as a means of improving worker and environmental health and safety. The TUR Act, a product of negotiations among citizen groups, business, and government over several years, has these goals: to decrease the amount of toxic waste generated in Massachusetts by 50 percent by 1997, to make toxics-use reduction the preferred method of compliance with other environmental laws and regulations, to promote reductions in the production and use of toxic substances in Massachusetts, to strengthen the enforcement of existing environmental laws and regulations, and to encourage coordination among state agencies administering toxics programs.

The TUR Act calls for firms using certain ("threshold") amounts of approximately 900 toxic chemicals to analyze and quantify their use of these substances and to undergo a detailed planning process aimed at identifying and evaluating options for reducing their use. Companies are not required to implement feasible options for toxics-use reduction; they must merely consider some options.

To maintain the quality of the planning process, a company's TUR plans must be certified by an individual who either has several years of experience in pollution-prevention activities at a company level or has completed a 48-hour training course. Such individuals must obtain periodic recertification through continuing education and training.

A company may count any of the following as toxics-use reduction:

• input substitution (i.e., substitution of one chemical for another to produce essentially the same product)
• product reformulation (i.e., alteration of the product)
• production unit redesign (i.e., using production equipment of a different design than previously used)
• production unit modernization (i.e., upgrading or replacing production equipment or methods)
• improved operations and maintenance
• closed-loop recycling and reuse.

A process change is not considered toxics-use reduction if it shifts toxic risks from one place to another (e.g., from the environment to workers). "Toxics use reduction," the act states, "means in-plant changes to production processes, that reduce, avoid, or eliminate the use of toxic or hazardous substances or generation of hazardous by-products per unit of product, so as to reduce risks to the health of workers, consumers, or the environment without shifting risks between workers, consumers, or parts of the environment."

The detailed TUR planning process is essentially a within-company corporate process, but it could be a public process if the general public were to be included in the planning team. One opportunity for the public to get involved arises when workers are involved (as is often the case), and when the particular workers are collaborating with citizen groups (as is sometimes the case).

The planning process generally does not question the manufacture of the product, but only how it is produced.

The planning process consists of the following steps:

• Setting goals and priorities. The company establishes a TUR team consisting of top management, environmental/health and safety officers, engineers, financial, sales, and legal departments, purchasing, and shop-floor employees. The team develops a corporate TUR policy statement, sets TUR objectives, and notifies workers of the Act and its requirements.

• Analysis of the process. The TUR team collects general data on chemical use and costs. Through a flow diagram process, the team divides the company into individual production processes and identifies how and why certain chemicals are being used and any sources of inefficiencies in use. The team also quantifies how much chemical is being used and how much hazardous waste is being produced in each production process per unit of final product.

• Identification of TUR options. The TUR team conducts an open "brainstorming" session of potential options to reduce toxics use in the facility, including options that may not seem currently feasible. The team may seek suggestions from additional shop-floor employees or outside assistance from vendors or technical assistance programs.

• Evaluation of TUR options. The TUR team screens out options that are not technologically feasible and conducts a two-step cost analysis of the remaining options. In the first step, the following costs are considered for each toxic chemical: Raw materials, labor, labeling, training on use, permits and fees, fines and penalties, waste disposal, monitoring and reporting, insurance, and possible litigation. A progressive company might estimate worker productivity and health costs. In general, however, environmental, worker health, and safety costs are calculated only as costs avoided (i.e., fines and penalties). Costs outside of the plant, e.g., to the community from polluted air, are not directly estimated. In the second step, total costs of each option are calculated over an investment lifetime, taking into account interest rates and tax credits. Finally, the team evaluates environmental, health, and safety differences among the options and selects the best option for implementation.

• Implementation of TUR changes. If the company wishes to implement a change it has identified and evaluated in earlier steps, it creates a timeline and secures approval and funding for the changes.

• Measurement of progress. An important part of the Act for the government, companies, and public is the measurement of progress in reducing toxics use and wastes. While companies are required to undergo the planning process every two years, each year they are required to quantify and compare their toxics use and wastes with their first year of reporting under the Act. This requirement helps to document successes and failures in the program.

Summaries of a company's full TUR plan, including information on toxic chemicals used, are publicly available. These summaries include

• required information on the quantities of the toxics a facility manufactures, processes, or otherwise uses
• environmental releases and by-products
• increases or decreases in toxics use
• a statement of company goals for the future
• any chemical substitutions that have been made
• toxics use reduction methods.

Information on how toxics are *used* provides a much fuller picture of the hazards to the public, to workers, and to the environment than does the information on direct releases to the environment required by the federal Toxics Release Inventory.[8] Under the TUR Act, companies must report the quantities of toxics they manufacture, process, or otherwise use.

For example, although a company may release relatively small amounts of a toxic chemical directly into the environment, large amounts may be incorporated into the product, creating hazards for the consumer, for workers in the plant, or (upon disposal) for the environment. Universal Forest Products, for instance, produces building products. The company's direct environmental releases and disposal (wastes) of two highly toxic metals, arsenic and chromium, as reported in 1992 under the Massachusetts TUR Act, were 5 pounds and 3 pounds, respectively. However, Universal shipped 176,000 pounds of arsenic and 196,000 pounds of chromium in its products.

Most of a TUR plan remains within the company and is not publicly available. However, the plan summary must be submitted to the Department of Environmental Protection, and it is then publicly available.

The public can use the summary information to pressure the company to adopt feasible TUR alternatives based on information about the amounts of toxics present in the product or released into the workplace or environment and on information about the company's use of toxics.

Companies differ in regard to how much information they supply in their summaries. Some companies (e.g., Polaroid) include a lot of information; some include very little. Therefore, the public may not get good information on all the options considered and rejected.

Neither the public nor the government can appeal an inadequate consideration of alternatives. However, the potential for pressure lies in knowing the amounts of chemicals used and the production unit involved. With this information, the public can demand more information directly from the company regarding options that were considered.

Companies covered under the TUR Act must pay fees based on the number of their employees and number of toxic chemicals over threshold amounts. These fees are used to fund the act's implementation and to fund two institutions that provide technical support, primarily to companies but also to the public: the Office of Technology Assistance and the Toxics Use Reduction Institute. The Office of Technology Assistance provides free assistance to companies in completing their plans and in identifying and implementing options. The Toxics Use Reduction Institute conducts research and training on toxics-use-reduction techniques and policies, funds university research and TUR projects, and is undertaking projects to increase citizen involvement in TUR policy and activities.

Each year the Massachusetts Department of Environmental Protection reports on the progress of the state's companies under the TUR Act. The 1998 report summarizes information from 530 "Large Quantity Toxics Users" and analyzes their toxic-use reduction from 1990 (the first year of reporting) to 1996. Adjusting figures to account for changes in production, the state reports that Massachusetts manufacturers reduced the amount of toxic chemicals they used to make their products by 24 percent, reduced the amount of wastes they generated by 34 percent, and reduced the amount of toxics they released to the environment by 73 percent in the period considered. Over this period, these companies collectively reduced their annual use of toxic chemicals by 209.8 million pounds, reduced their annual generation of hazardous waste by 37.5 million pounds, and reduced their annual releases of toxics to the environment by 15.1 million pounds (MDEP 1998).

The Montreal Protocol on Substances That Deplete the Ozone Layer

The Montreal Protocol on Substances that Deplete the Ozone Layer is an international agreement[9] to phase out most human uses of substances that destroy the ozone layer. Various ozone-depleting substances have been used for a variety of purposes: chlorofluorocarbons in refrigerators, foams, solvents, and aerosol products; halons in fire extinguishers; carbon tetrachloride as a feedstock for pharmaceutical and agricultural chemicals and as a solvent when chlorinating materials such as rubber. A number of committees are working to identify what alternatives exist or could be developed to eliminate the use of the ozone-depleting substances.

So far, the process has been working moderately well. In the case of chlorofluorocarbons, for instance, alternatives have been found more rapidly than planned, and the phaseout of their use will be complete in industrialized countries by 1996; a mere 50 percent phaseout by the year 2000 was originally scheduled by the Montreal Protocol signatory countries.

There is no question that the Montreal Protocol process could work even better and faster if each signatory country had a more profound commitment to environmentally sound and socially appropriate alternatives to ozone-depleting substances. The substitution of other, weaker ozone depleters (e.g., hydrofluorocarbons and hydrochlorofluorocarbons) for CFCs in refrigerators, for instance, has reduced, but continued ozone depletion.

The Montreal Protocol process and the World Bank (the primary implementing agencies of the Montreal Protocol ozone depleter phaseouts) have not emphasized development of small-scale technologies appropriate for developing countries. Of the $32 million allocated by the World Bank to refrigeration projects, for example, none are devoted to environmentally friendly technologies (Greenpeace 1994b). Greenpeace had to implement a strong advocacy campaign just to gain industrial, World Bank, and governmental recognition of "greenfreeze" refrigerators. These refrigerators use the hydrocarbon gases butane and propane, rather than HFCs, as coolants (Greenpeace 1994a). Likewise, while the hydrocarbon cyclopentane has been recognized by World Bank advisors as "the most cost-effective" zero-ozone depleting substance, all of the World Bank's grants to the domestic refrigeration foams sector go to CFC and HCFC technology (Greenpeace 1994b). This has also allowed DuPont chemical company to retain control of its CFC market advantage for CFC substitutes. By 1989, DuPont was planning to build a $20 million HCFC-123 plant (ENDS 1989).

In addition, loopholes allow the United States to continue production of CFCs until 2005 for "essential use and export" and an illegal, "black" market for CFCs is continuing illegal use (Vallette 1995). However, we are further along globally toward phaseout of ozone-depleting substances than we would have been without phaseout goals that require consideration and development of alternatives.

From 1992 to 1997, I served as a member of the Methyl Bromide Technical Options Committee. This Montreal Protocol Committee was charged with assembling a report on existing and potential alternatives to using the potent ozone-depleting pesticide, methyl bromide, as a soil fumigant (i.e., to kill all soil organisms with poisonous fumes); structural fumigant (e.g., in ships' holds, granaries, and homes); and fumigant of durable and perishable commodities of international trade such as wood, grains, fruits, vegetables, and flowers.

In other words, the Committee was charged with conducting an alternatives assessment. Its report would be used to help determine the schedule of a worldwide phaseout of methyl bromide use.

The process of exploring alternatives to methyl bromide within the Methyl Bromide Technical Options Committee was eye-opening. Only a small minority of the 52-member committee was publicly supportive of alternatives to methyl bromide, and even fewer knew of non-chemical alternatives to the pesticide. However, the Committee's mandate to discuss all feasible technical options to the use of methyl bromide lent clout to the minority of members who did have knowledge of non-chemical alternatives. If a committee member could provide evidence that a particular alternative was in use or close to availability, the alternative had to be entered into the report.

The definition of an alternative to methyl bromide, adopted by the Methyl Bromide Technical Options Committee early in its proceedings, also turned out to be critical: "A technology demonstrated or in use in one region of the world that would be applicable in another unless there were obvious technical constraints to the contrary (e.g. a very different climate)." This definition was reiterated at the start of every Committee meeting after the 1993 meeting in Nairobi (Banks 1995).

Notably, economic considerations were technically not the realm of the Committee; they were left to the Economics Committee. The Technical Options Committee thus was left free to discuss a full range of options. The resulting picture of alternatives, including non-chemical alternatives, is impressive. The Committee revealed that technically, alternatives already exist for more than 90 percent (by volume) of worldwide methyl bromide use (MBTOC 1994).

The U.S. Department of Defense, for instance, together with shipping companies and the University of California at Davis, has been successful in developing an alternative to one use of methyl bromide. Prior to 1994, the Department of Defense had been shipping fresh produce by air to U.S. military personnel stationed overseas. This produce was fumigated with methyl bromide to quickly kill organisms (e.g., insects or mites) that are prohibited (quarantined) in the country to which the produce was being shipped.

The alternative the Defense Department developed is "controlled atmosphere" technology in refrigerated vans (DSR PAC 1994). Controlled atmosphere technology uses optimum combinations and concentrations of nitrogen, carbon dioxide, and oxygen to retain freshness of perishable commodities and to eliminate or prevent the presence of quarantined organisms. The produce is treated during shipment by sea, but it arrives in better condition than when it had been fumigated and flown. The Department of Defense is saving millions of dollars by using controlled atmosphere technology (Lt. Commander Robert B. Gay, MSC, U.S. Navy, letter to Mary O'Brien, March 14, 1994). While developed for shipping produce to U.S. military personnel, the technology will be applicable to many fruits and vegetables traded between countries.

Many committee members resisted reporting the existence of alternatives. These members were invested one way or another (either economically or in a government agency supportive of methyl bromide-using industries) in continuing the use of methyl bromide. If the Committee had not had such a clear mandate (i.e., to report on all known technical alter-

natives and potential alternatives to the use of methyl bromide) and definition of "alternative" (i.e., a technology demonstrated in one region of the
world is presumed to be applicable worldwide), the report would have
claimed that continued methyl bromide use was essential.

The definition of "alternative" was also critical after the Committee
issued its report stating that alternatives exist for more than 90 percent of
worldwide methyl bromide use. Several Committee members claimed that
such a wide range of alternatives aren't in use everywhere (Banks 1995).
The Chair of the Methyl Bromide Technical Options Committee was able
to point to the definition and defend the "90+ percent" claim with examples recorded in the report.

Public reporting of the existence of alternatives (as in the case of methyl
bromide) is a powerful process. It can be extremely threatening to the
status quo.

The International Joint Commission on Great Lakes Water Quality: Assessment of Alternatives to the Chlorine Industry

The following chronology of one commission's progress, over decades,
traces its deepening understanding of the nature of threats to a particular
water body and the need for alternative societal behaviors if toxic contamination of the water body is to be virtually eliminated. The Commission
eventually came to the conclusions that chemical-by-chemical risk assessment was not going to be adequate to address the toxic threats and alternatives assessments were in order.

In 1909 the governments of Canada and the United States formed the
International Joint Commission (IJC) to oversee the Boundary Waters
Treaty, which guides the behavior of the two countries in regard to the
Great Lakes.

Over the decades, evidence grew that persistent toxic substances were a
major cause of deterioration of the water quality of the Great Lakes and
the health of wildlife associated with the lakes. These persistent toxic substances were causing visible effects on wildlife, including reproductive failure, birth defects, nervous-system damage, chick fatalities, eggshell
thinning, and fish cancers.

The 1978 Great Lakes Water Quality Agreement, signed by the U.S.
and Canadian federal governments, insisted that "the discharge of toxic

substances in toxic amounts be prohibited and the discharge of any or all persistent toxic substances be virtually eliminated."

The Great Lakes Water Quality Agreement defines a toxic substance as one "which can cause death, disease, behavioral abnormalities, cancer, genetic mutations, physiological or reproductive malfunctions or physical deformities in any organism or its offspring, or which can become poisonous after concentration in the food chain, or in combination with other substances."

The 1991 report of the IJC's Virtual Elimination Task Force (VETC 1991) is a thoughtful alternatives-assessment report on how the nations, states, communities, and industries bordering the Great Lakes can virtually eliminate the discharge of persistent toxic substances to the Great Lakes. It notes, for instance, that the U.S. Toxic Substances Control Act provides the authority to ban, phase out, and restrict uses of toxic chemicals. It suggests determining what "right price" means in terms of encouraging virtual elimination of a toxic chemical, as well as targets and timelines to evaluate process. It emphasizes using taxes on discharges of persistent toxic substances to encourage and support development of alternatives to the use of those substances.

In its *Sixth Biennial Report on Great Lakes Water Quality* (1992) the IJC recommended a definition of a persistent toxic substance: "all toxic substances with a half-life[10] in any medium—water, air, sediment, soil or biota [plants and animals]—of greater than eight weeks, as well as those toxic substances that bioaccumulate[11] in the tissue of living organisms." In this same document, the IJC rejected a risk-assessment approach to persistent toxic chemicals: "We conclude that persistent toxic substances are too dangerous to the biosphere and to humans to permit their release in *any* quantity" (IJC 1992). This rejects the risk assessment approach because it rejects the process of estimating how much of any, persistent toxic substance would be insignificant.

The IJC also noted that industrial uses of chlorine need to be phased out, because chlorine is the source of so many persistent toxic substances entering the Great Lakes. One of the *Sixth Biennial Report*'s thirteen recommendations is that "the Parties [Canada and the United States], in consultation with industry and other affected interests, develop timetables to sunset the use of chlorine and chlorine-containing compounds as industrial feedstocks and that the means of reducing or eliminating other uses be examined."

In support of this recommendation, the IJC noted that of 362 chemicals of concern confirmed to be present in the water, sediment, or living organisms of the Great Lakes, approximately half are synthetic chlorinated organic (carbon-containing) substances. The IJC further noted that even though many of the thousands of chlorinated organic substances that are produced have not been identified or proven to be individually toxic, "it is likely that many of these chemicals—because of their chemical characteristics—will be identified as persistent toxics and hence substances to be virtually eliminated and subject to zero discharge" (IJC 1992).

In this report, then, the IJC rejected a chemical-by-chemical risk-assessment approach to each of thousands of organochlorines, saying that they are likely to be toxic as a class, and therefore their production and use must be phased out as a class.

In 1993 the Virtual Elimination Task Force published *A Strategy for Virtual Elimination of Persistent Toxic Substances*, outlining a general strategy, illustrated by clear flow charts and accompanied by case strategies for the virtual elimination of mercury, PCBs, and chlorine. One of the case strategies, titled "Case Study. Application of a Virtual Elimination Strategy to an Industrial Feedstock Chemical—Chlorine" (Muir et al. 1993), presents a "use tree" of all known industrial uses of chlorine (figure 11.1), and then outlines the availability of alternatives and the prospects for sunsetting most uses.

All of the above reports are available from the International Joint Commission, 100 Ouellette Avenue, Windsor, Ontario N9A 6T3. They are useful models of alternatives assessments and arguments against risk assessment when a big picture of damage is understood.

Pesticide Use Reduction in Denmark, the Netherlands, and Sweden: Assessments of Alternatives to Pesticides

Denmark, the Netherlands, and Sweden have each committed to specific numerical targets and timetables for reduction in the use of pesticides (Hurst 1992):

Denmark: 50 percent reduction in active ingredient use by 1997.

Netherlands: 50 percent reduction in active ingredient use by 1997; 38 percent reduction of emissions to air, and 70 percent reduction of emissions to surface water by 1995.

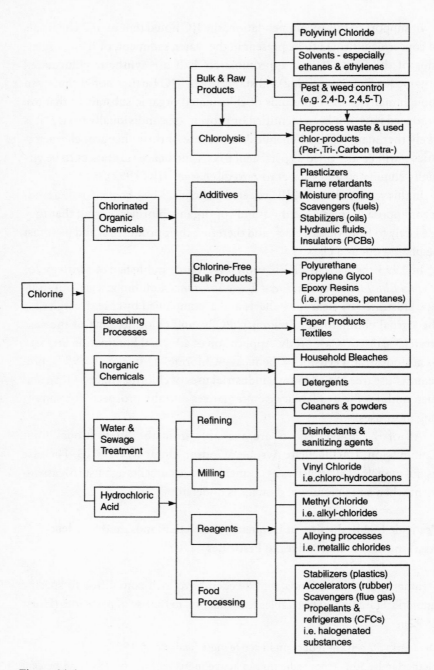

Figure 11.1
A simple chlorine-use tree that identifies major bulk products, processes, and application categories. Source: Muir et al. 1993.

Sweden: 50 percent reduction in active ingredient use 1986-1991 (essentially met); further 50 percent reduction 1991-1997.

A number of social processes are being used to provide incentives for pesticide reduction, (such as pesticide re-evaluation, levies and taxes, and subsidies for conversion to integrated farming). Some of these relate directly to assessment of alternatives. Sweden, for instance, is funding research into the benefits of pesticide-free strips in fields and promoting integrated pest management (IPM).[12] Extension services to farmers, emphasizing IPM, are being increased in all three countries. Denmark is taxing pesticides 3 percent to fund environmental research projects.

A key to the generation of alternatives to pesticides in the Nordic countries is that clear numerical goals have been set for reduction in pesticide dependency. However, actual measurement of progress toward the goals can be difficult, because some of the goals can be met simply by using more potent pesticides at lower volume or in fewer applications.

In the United States, the pesticide user and manufacturer industries resist and work to defeat proposals to establish numerical pesticide-reduction goals. They emphasize the economic benefits produced by chemical pest control and maintain that existing regulation is sufficient to detect and respond to any adverse environmental health impacts. Occupational health and environmental advocates, on the other hand, argue that current monitoring systems and risk assessment are incapable of tracking, characterizing, or prohibiting all the problems caused by pesticide use (Pease et al. 1996).

When we decided to place a human on the moon, we did it. Likewise, we can set equally clear goals to restore the environment, significantly reduce toxics use, increase roadless habitat, or reverse population growth. Once goals are set and policies are developed that provide incentives to meet the goals, creative people will develop and implement alternatives.

The Challenge: Making Alternatives Assessment Commonplace

Alternatives assessment may be a common-sense approach to decision making, but it is not commonplace. The main reason it is not commonplace is that alternatives assessment challenges the status quo, in which business as usual continues practices that damage the environment. Two brief examples of avoidance of alternatives assessment illustrate this.

Example: Resistance to Considering Alternative Sites for a Telescope
The University of Arizona (as leader of a consortium of astronomers) want-
ed to establish a series of telescopes on Mount Graham in the Pinaleño
Mountains, even though doing so might cause the extinction of the Mount
Graham red squirrel. Initially, the university did not want to discuss alter-
native mountain ranges where the astronomy facilities might be built. Under
NEPA, however, it had to discuss alternatives in an Environmental Impact
Statement. In order to skip this step, the university, using public money, successfully lobbied Congress (through a Washington, D.C. lobbying firm) to
exclude the project from NEPA (Warshall 1994).[13]

Example: Resistance to Alternatives to Methyl Bromide Use
The U.S. Department of Agriculture maintains that U.S. agribusiness needs
to use the ozone-depleting pesticide methyl bromide. The department has
actively sought change of the U.S. Clean Air Act requirement that methyl
bromide use be phased out by 2001, and has worked to prevent the establishment of a rapid international phaseout schedule through the Montreal
Protocol process.

A member of the United Nations Methyl Bromide Technological Options
Committee, Ken Vick of the U.S. Department of Agriculture, vigorously
protested inclusion in the Committee Report of the accurate statement that
"nine existing alternatives [to methyl bromide] are approved for quarantine
[import] purposes with perishables [e.g., fruits, flowers]." Vick supported
his objection on the ground that "a little information is a dangerous thing."

In other words, a little public information about alternatives is dangerous to the U.S. status quo of requiring that perishable commodities be fumigated with the ozone-depleting pesticide methyl bromide to meet U.S.
quarantine regulations.

When citizens see that there are reasonable, feasible alternatives to
destroying the world, some of them will exert political pressure to implement at least some of those alternatives. They may press individual facilities to adopt alternatives voluntarily. They may press federal agencies to
pursue particular environmental policies. They may press state agencies to
require proven technologies that had earlier been presented as "infeasible." They may press courts to pay attention to legal mandates to consider
alternatives.

The more that facilities, agencies, communities, states, and nations undertake alternatives assessments, the more skilled they will become at assessing alternatives, and the more assessment of alternatives will be seen as "the decent way" to operate in the world.

Finally, the mere statement by an industrial facility or corporate farm that it could behave better highlights the fact that corporations and facilities are choosing to regard air, water, soil, wildlife, and our bodies as cheap sources of materials and/or as free depositories for their toxics waste. Alternatives assessment makes clear that there are alternatives to treating the Earth and its inhabitants as endless sources of materials and bottomless waste receptacles.

It may be that industrialized societies will eventually do an alternatives assessment of the laws that allow "private" business to invade public arenas (air, water, soil, wildlife, our bodies) when alternatives exist to avoid doing so.

But first, we have to get those alternatives on the table.

12

We Know How to Push for Alternatives Assessments

Some citizen groups and Native American tribes have opposed risk assessment and have urged alternatives assessments.

Risk assessment of bad options is the main decision-making process in our society at the moment. Unfortunately, too many citizen groups think they have to cooperate with it, but they don't have to. When they do resist risk assessment, citizen groups employ two major strategies: saying "No" to unnecessary bad options or to risk assessment itself and advocating for "Yes" to better alternatives.

The "No" strategy answers the question "How much damage shall we allow through X activity?" with "None; don't do that activity." Sometimes activists using this strategy accomplish a simple "No" without installing some parallel, new alternative, as when they block the permitting of a cement kiln to burn hazardous waste and the cement kiln simply continues burning the fuel it had been burning.

Activists are likely to choose a "No" strategy when they are conscious of their right not to be trespassed upon by dangerous substances, or when they are conscious of their right and responsibility to defend the habitat and continued existence of others.

David Ozonoff (1993) relates a story about a surgeon named John Snow who used the "No" strategy in 1853:

It was just 140 years ago that a London surgeon, John Snow, first plotted on a map the occurrence of new cases of cholera,[1] then raging in epidemic form in his city. An extremely astute observer, he noted that many of the cases clustered around the Broad Street [drinking water] pump in Golden Square. Unable to get the cooperation of municipal authorities, so the legend goes, he literally stopped the epidemic single-handedly by taking the handle off the pump, thus preventing further exposure.

Snow's action was a simple "No" to the continuance of the cholera epidemic, which, he concluded, was preventable.

The "Yes" strategy bypasses the question "How much damage shall we cause through X activity?" by saying "Here is a much better and yet feasible alternative activity." The more determined players of this strategy do not easily concede to charges of political infeasibility. They know that most significant environmental and public-health advances have been considered "politically infeasible" at one time or another. They likewise reason that if we can undertake the intense Manhattan Project to make nuclear bombs, or a space effort to transport humans to the moon, then we can institute alternatives to destroying the environment.

The "Yes" strategists reckon how best to get alternatives onto the table that are otherwise being shoved under the rug.

Of course, assessment of potential damages ("risks") plays a part in both the "No" and "Yes" strategies. After all, the reason for saying "No" is that the activity or proposal being opposed has been assessed by at least some people as dangerous, and the reason for saying "Yes" to an alternative is that its potential damages appear to be significantly less than those associated with the proposed or ongoing activity. Both strategies therefore involve risk assessment. But when alternatives are on the table, the technicalities of quantitative risk assessment are likely to be overshadowed by more fundamental questions raised by the various alternatives, such as the following:

• Who bears the costs of and who profits from the different alternatives?
• Which alternatives can be more effectively watchdogged and controlled by affected citizens?
• Which alternatives can be sustained long into the future?

Several citizen groups have opposed risk assessment and demanded alternative approaches.

Greenpeace: Opposition to Risk Assessment

The organization called Greenpeace, a leader worldwide in opposing risk assessment, has attacked the "assimilative capacity" assumption of risk assessment, namely that risk assessors can know how much of a given harmful activity can be withstood (assimilated) by the environment.

Greenpeace cites the existence of an alternative approach: use of the "precautionary principle." In 1990, Greenpeace submitted a four-page paper to the parties (i.e., government representatives) to the London Dumping Convention (1972), whose signing nations develop international agreements governing sea dumping. In this paper, titled Protection of the Marine Environment through the "Precautionary Action" Approach, Greenpeace asserted the following:

Where once there was little or no environmental policy, its evolutionary development has followed an approach based on "allowable" emissions or discharges. This approach is often called the "assimilative capacity" approach to pollution because it is based on the assumption that the environment has a capacity to receive, and render harmless, the vast quantity and variety of industrial inputs.

This traditional "permissive" approach does not represent a sound scientific approach to the protection of the environment. The existing body of scientific literature makes it clear that even the most sophisticated environmental impact assessment models contain substantial inherent uncertainty due to the overwhelming diversity and complexity of biological species, ecosystems, and chemical compounds entering the environment. What were once considered perfectly safe levels of particular inputs into the environment subsequently have been determined unsafe. The legacy of environmental degradation attests to this fact.

Greenpeace then listed numerous international commissions, councils, and conventions and the European Community Parliament as settings in which the "precautionary action" approach had been adopted. It cited the Nordic Council's 1989 Conference on Pollution of the Seas as one definition of the precautionary principle:

The need for an effective precautionary approach . . . [of] eliminating and preventing pollution emissions where there is reason to believe that damage or harmful effects are likely to be caused even where there is inadequate or inconclusive scientific evidence to prove a causal link between emissions and effects.

Greenpeace promotes "clean production" as a means to implement the precautionary principle. Clean production employs manufacturing, transport, disposal, and use activities that drastically reduce or eliminate use of toxic chemicals (and subsequent release of toxic chemicals into the environment). The concept of clean production, according to Greenpeace, also necessarily questions the need for production at all in certain cases: "In the first place, a Clean Production Approach questions the very need for the product or looks at how else that need could be satisfied or reduced." (Kruszewska 1995)

Greenpeace has followed through on its commitments to clean production and the precautionary principle in its various campaigns, thereby changing the nature of each debate. For instance, Greenpeace has

• helped develop the precautionary principle-based UN resolution to ban drift-net fishing on the high seas (United Nations 1990)

• worked with African governments to produce the Bamako Convention, which prohibits imports of banned or unregistered pesticides from outside of Africa and greatly restricts imports of hazardous wastes into Africa (Organization of African Unity 1991)

• worked with the International Joint Commission on Great Lakes Water Quality to call for the phaseout of industrial uses of chlorine (IJC 1994)

• worked with German and British refrigeration engineers to develop butane- and propane-based refrigerators that avoid the use of hydrochlorofluorocarbons and hydrofluorocarbons, which are ozone-depleting substitutes for the more potent ozone-depleting chlorofluorocarbons (Greenpeace 1994a).

Greenpeace opposes risk assessment of dangerous options, and works to bring alternatives to the table.

Greenpeace, the International Joint Commission, and the American Public Health Association: Opposition to Risk Assessment of Chlorinated Chemicals

Production, use, and disposal of chlorine and chlorinated organic (i.e., chlorine- and carbon-containing) chemicals inevitably result in releases into the environment of numerous poisonous and persistent chlorinated organic chemicals. Many of these are stored in the fats of animals and accumulate at higher and higher concentrations in animals that are higher and higher in the food chain.

Greenpeace has served as the leading international non-governmental organization calling for an end to industrial and consumer use of chlorine. It has undertaken numerous chlorine-related campaigns, including the "Chlorine-free in '93" campaign begun in 1988, a campaign to get *Time* magazine to use chlorine-free paper, efforts to get dry cleaning establishments to use water steam rather than perchlorethylene, and cooperation with refrigeration specialists to develop butane- and propane-based refrigerators that would avoid the use of chlorinated fluorocarbons.

On occasion, Greenpeace has commented on risk assessments. In 1988, for instance, at Greenpeace's request, Carol Van Strum wrote a rebuttal to an industry-influenced attempt by the U.S. EPA to lower its dioxin standards (Van Strum and Merrell 1988). The EPA proposed to average a range of risk-assessment estimates of dioxin's toxicity. This proposal was eventually abandoned. In 1994, Greenpeace Senior Scientist Pat Costner debunked the EPA's risk-assessment assumption that dioxin incinerators destroy 99.9999 percent of dioxin (Costner 1994; see chapter 2, example 3), and other chlorine-based discharges. However, Greenpeace has taken apart these risk assessments only as part of its call for implementation of the precautionary principle and an end to the use of chlorine (Thornton et al. 1993). Greenpeace has not debated how much exposure to chlorinated compounds is acceptable. Rather, it has consistently pointed to the existence of developed alternatives for almost all uses of chlorine.

The International Joint Commission for the Great Lakes, in recommending that chlorine and chlorine-containing compounds be phased out as industrial feedstocks, noted that numerical water-quality "standards" based on risk assessment are "irrelevant for persistent toxic substances" (IJC 1994), which accumulate in biological systems and substrates. (See chapter 11.)

In November 1993 the American Public Health Association passed a resolution concluding that "there should be a rebuttable presumption [i.e., an assumption that someone would have to prove otherwise with data] that chlorine-containing organic chemicals pose a significant risk" (APHA 1993). In passing this resolution, the APHA relied in part on the International Joint Commission's Science Advisory Board's conclusion (IJC 1991) that "policies to protect public health should be directed toward eventually achieving no exposure to chlorinated organic chemicals as a class rather than continuing to focus on a series of isolated, individual chemicals" (ibid.).

All three of the above-mentioned organizations have called on public bodies to require non-chlorinated technologies as alternatives to chlorine use.

NC WARN and the Chemical Weapons Working Group: Alternatives to Incineration Risk Assessment

Numerous citizen groups and coalitions have worked fiercely to defeat proposals for incinerators that would burn hazardous waste.

NC WARN, a coalition of citizen groups in North Carolina, has defeated 20 attempts to site a hazardous-waste incinerator in that state. The coalition's strategy has relied on the slogans "Not here, not there, not anywhere" and "No compromise, no negotiations."

NC WARN has also worked to replace incinerator siting efforts with toxics-use-reduction policies. The General Assembly of North Carolina eventually abolished the Governor's Waste Management Board, created the Pollution Prevention Advisory Council, and cut the appropriations of the North Carolina Hazardous Waste Management Commission (incinerator siting commission) to zero (NC WARN, unpublished paper).

The Chemical Weapons Working Group is a coalition of citizen groups that oppose incineration of stockpiled military chemical (nerve gas) weapons. It is a project of the Kentucky Environmental Foundation of Berea, Kentucky.

Chemical weapons are of two basic types: (1) organophosphate nerve agents and (2) vesicant agents, which cause blistering and death of tissues following contact. Since 1984, citizens living near sites where the U.S. Army stores its chemical weapons have been opposing proposals to incinerate these weapons. Citizens are concerned with toxic releases from chemical-weapons incinerators and with the likelihood of severe accidents, which could release chemical-weapons agents and other toxic chemicals into the air. The Army's chemical-weapons incinerator on the Johnston Atoll in the South Pacific, for example, has experienced a multitude of shutdowns and accidents and has been subject to large budget overruns (Picardi et al. 1991).

The citizens have been pressing the Army to utilize more benign methods of disposal, including drainage of the nerve agent from weapons and neutralization or other acceptable treatment of the chemical agent (ibid.). Neutralization in this case refers to a process in which the toxic agent is rendered non-toxic through a chemical reaction. Different reactions are feasible for different chemical agents, and combinations of techniques can be used.

Other possible non-incineration treatments include steam gasification (which would detoxify the nerve agent by mixing it with steam at high temperatures), chemical and thermal treatment with molten metals or molten salt, breaking down toxic chemicals by a combination of moderate temperature and high pressure (i.e., "supercritical water oxidation"), using bacteria or enzymes to break down the nerve agent (i.e., "biological methods"), and using potassium or sodium hydroxide and catalysts to chemically neutralize mustard agent. Many of these are closed-loop methods requiring little or no discharge of contaminants into air or water (CWWG 1994; for an additional method, see Picardi et al. 1991).

The Chemical Weapons Working Group has traced the internal politics of the Army as it opted for incineration over neutralization, has followed recent promising laboratory tests of chemical neutralization and other disposal technologies, and has been making its position clear to Congress and to other decision makers.

In addition to its incinerator in the South Pacific, the Army has an incinerator in Tooele, Utah. It now has built an incineration bureaucracy, and numerous specialists have spent good parts of their professional careers promoting incineration. Meanwhile, the Army has largely ignored its own research experience with neutralization.

Primarily as a result of citizen resistance to incineration and citizen pressure to consider alternatives, the National Research Council is currently re-examining the technological feasibility of alternatives to incineration. It had examined alternatives earlier, in 1984, when it recommended incineration on the basis that chemical methods were too slow and that they produced hazardous waste (NRC 1984). At that time, the toxic-emission and waste-generation problems of incineration were not fully understood (Picardi et al. 1991).

Additionally, in 1996, Congress agreed to fund the evaluation and demonstration of at least two alternative technologies to incineration (Public Law 104-208; Brooke 1997). The Department of Defense, acknowledging citizen participation, has established a Dialogue on Assembled Chemical Weapon Assessment, which includes representatives of national activist organizations that regularly work on this issue, Native American tribes, and affected community members. This group has been assisting the department to assess alternative technologies (Keystone Center, undated).

The Confederated Tribes of the Umatilla Indian Reservation: Alternatives to Nuclear Waste Risk Assessment

Dump sites of persistent, hazardous materials make it extremely difficult to avoid decision making based on risk assessment. The response of citizen groups cannot feasibly be "Clean it to pristine conditions," because realistically the sites will be contaminated for long, long periods of time. The situation necessarily involves asking "How much pollution will we allow to remain?" Risk assessment is used to provide an answer.

One of the most polluted hazardous-waste dumps on Earth is the 570-square-mile Hanford Nuclear Reservation, adjacent to the Columbia River in south-central Washington. It is managed by the U.S. Department of Energy. Hanford's chief mission has been to produce plutonium for use in nuclear weapons. Hanford's "B Reactor" was the first plutonium-production reactor in the world. Plutonium created within this reactor fueled the first atomic explosion and formed the core of the bomb that exploded over Nagasaki. The reactor was built in less than a year. Hanford contains nine production reactors and four chemical separation plants.

Over the period 1944–1987, Hanford Nuclear Reservation became the main U.S. site for nuclear fuel fabrication, irradiation, and chemical separation.[2]

At Hanford, spent nuclear fuel exists in an obsolete facility a few hundred yards from the Columbia River. Over time, one basin leaked millions of gallons of contaminated water, which reached the groundwater in 1997 (Raloff 1997).

Five chemical separation buildings and their underground tanks exist on Hanford; such facilities are among the most radioactive places in the United States.

In addition, large amounts of toxic chemicals were used, stored, and dumped at Hanford. At the moment, underground plumes heavily contaminated with chromium are flowing into the Columbia River from beneath the Hanford Reservation (PNL 1997; Columbia River United, undated). Chromium is highly toxic to fish.

Ceded lands of the Confederated Tribes of the Umatilla Indian Reservation (the "Umatilla Tribes," including the Walla Walla, the Umatilla, and the Cayuse) include a portion of the Hanford Nuclear Reservation. As a government separate from the United States, these tribes retain treaty rights within the Hanford Nuclear Reservation, including rights to hunt, gather, and fish on the land and in the streams. These traditional, treaty-protected activities cannot be carried out on land and in rivers that are so polluted that they do not support indigenous vegetation and wildlife.

The Umatilla Tribes are showing that conventional risk assessments of Hanford wastes do not address the tribes' activities and culture. In 1995 they prepared a document titled Scoping Report: Nuclear Risks in Tribal Communities (CTUIR 1995). Box 12.1 summarizes some of the elements the tribes say must be included in assessments of various alternatives for dealing with the radioactive and toxic wastes on the Hanford Nuclear Reservation.

It might seem that the considerations in box 12.1 could be plugged into a standard risk assessment, but they cannot. For instance, the tribes' beliefs that every impact on the environment is an impact on tribal members (item 12) would change the entire process of risk assessment; wildlife and human impacts could not be considered separate issues. Likewise, if the full range of possible cleanup and restoration activities called for (item 9) are included, the risk assessment becomes an alternatives assessment.

In the Scoping Report, the Umatilla Tribes cite favorably two examples of integrated, holistic environmental planning that have taken place at the Hanford nuclear site[3]:

• the Hanford Future Site Uses Working Group, which came to consensus on nine recommendations, including immediate protection of the Columbia River, forceful action on groundwater contamination (e.g., immediate pumping out of contaminants such as chromium), and involvement of the public in future planning.
• the Hanford Tank Waste Task Force, which called for development of a set of technological alternatives for dealing with buried, leaking tanks of toxic chemicals and radioactive materials. The Task Force developed ten principles by which to select action alternatives.

Based on these smaller efforts, the Umatilla Tribes write:

The development of clearly defined principles, goals, and decision criteria and a single sitewide engineering design basis which directly incorporates values, expectations, interests, and rights will be essential to provide the holistic framework necessary for both technically defensible and politically acceptable decisions. This process must include the fundamental establishment of a comprehensive and effective intergovernmental process built together with tribal sovereigns, and not just in response to them.

In this paragraph, the Umatilla Tribes assert that their Native American rights and values must be considered and that the plans for Hanford must be not only technically sound, but the result of real collaboration between

Box 12.1
Elements to Be Assessed in Environmental Assessments of Nuclear Sites

Environmental assessments of the Hanford Nuclear Reservation and any other Department of Energy (DOE) site must address the following:

Unique and multiple use of treaty-reserved rights and resources for subsistence, ceremonial, cultural, and religious practices[1]

Multiple exposure pathways that result from cultural resource use (hunting, gathering, fishing) that are not considered in typical "suburban" exposure scenarios

The reality that tribal communities' lifestyles result in disproportionately greater than average exposure potential, although the size of the populations may preclude epidemiological statistical power

Cumulative effects of multiple chemicals/paths/sites (e.g., past, present, and future impacts of Hanford Nuclear Reservation, Umatilla Army Depot, and Boardman Bombing Range; past, present, and future impacts on the resources by regional non-nuclear activities)

Time and risks to future generations

What is truly desirable and achievable (i.e., benefits of remediation and restoration)

Environmental justice and the right to a safe and healthful environment

Qualitative, intangible well-being (aesthetic, physical, economic, community and future), peace of mind, and sustainability

A full range of alternatives for proactive remedial and restoration actions

Information gathered from multiple sources of knowledge. This alone requires tribal[2] and public participation.

Retention of future natural resources potentials and options

The religious significance of all indigenous plants and animals and their interdependent relationships and interactions

Injuries, as well as "risks"

based on CTUIR 1995

---

1. Exercise of these rights depends on the health of the natural resources, and activities associated with these rights are as important as material products.

2. As a sovereign government, the Confederated Tribes of the Umatilla Indian Reservation (CTUIR) is an entity with rights apart from the public. In addition, the CTUIR is a Trustee for Natural Resources under the federal Comprehensive Environmental Response, Compensation and Liability Act (CERCLA).

two sovereign, culturally different entities. Moreover, they declare that biological factors, which include only risks of death or illness, are not the only factors that must enter into assessments. Instead, the assessments must include all the elements seen in box 12.1.

The Hells Canyon CMP Tracking Group: Alternatives to Risk-Based Management

Risk assessment of the effects of activities on ecosystems is called *ecological risk assessment* (USEPA 1992). Like chemical risk assessment, ecological risk assessment is used to estimate how much of particular human activities an ecosystem can "assimilate." (Ecological alternatives assessment, on the other hand, involves considering the degree to which humans can adapt their ways so as to live within the needs of the ecosystem. See O'Brien 1995.)

The following is the story of a citizen coalition's effort to improve the U.S. Forest Service's management of Hells Canyon by proposing an ecosystem-friendly alternative for an upcoming Forest Service alternatives assessment. The coalition wants the Forest Service to manage human activities so they adjust to the needs of Hells Canyon, rather than to estimate how much human activity Hells Canyon can assimilate.

The Hells Canyon National Recreation Area (NRA), located in northeastern Oregon and western Idaho, includes the Hells Canyon Wilderness and the Snake River. It is managed by the Wallowa-Whitman National Forest (NF). Hells Canyon is 8000 feet deep from the alpine peaks of the Seven Devils Mountains in Idaho to the Snake River. It includes pine, larch, and Douglas fir plateaus and canyons and native bunchgrass ridges and slopes. The wildlife includes bears, elk, bighorn sheep, mule deer, lynx, cougar, bald and golden eagles, chinook salmon, and sturgeon.

In 1994 the Wallowa-Whitman NF announced that it would be devising a new management plan (called a Comprehensive Management Plan, or CMP) for the Hells Canyon NRA. The last such plan for Hells Canyon had been prepared in 1979. The plan was to be prepared as an Environmental Impact Statement (EIS) under the National Environmental Policy Act (NEPA) and was to include "all reasonable alternatives" (Council on

Environmental Quality 1992a). This meant that the EIS would have to describe all reasonable alternatives for management of the Hells Canyon NRA.

Early in 1995, the Wallowa-Whitman NF invited the Hells Canyon CMP Tracking Group to develop an alternative to its EIS. The Tracking Group is made up of representatives from fifteen environmental and hunting organizations, the Nez Perce tribe, and the Confederated Tribes of the Umatilla Indian Reservation. (Hells Canyon is the ceded lands of the Nez Perce.) The tracking Group had convened in early 1994 to track the CMP process.

In order to comply with the Forest Service framework, the Tracking Group's alternative had to propose management of numerous activities in Hells Canyon within a Forest Service framework of goals, objectives, standards, and guidelines, addressing the following topics:

access (i.e., roads)
recreation
aquatic conservation
Native American and non-Native American cultural sites
forest, riparian,[4] and grassland wildlife habitats
geology
"unique biological habitat"
scenery
science/education
soils
wildlife.

These and other topics involve numerous human activities that affect the environment (e.g., livestock grazing, logging, jet boats, off-road vehicles, drive-through tourism, and snowmobiles).

The Tracking Group designed all proposals in its Native Ecosystem Alternative around four principles, each of which stands in contrast to risk-based principles employed by the Forest Service:

• subordination of human activities to ecosystem protection (which is required by the Hells Canyon NRA Act) and recovery
• Forest Service decision making for no-impact or least-adverse-impact alternatives for human activities, whether Forest Service activities or recreational or commodity uses (i.e., logging, livestock grazing)

• standards for activities that are measurable by the public and based on ecosystem protection and maximum feasible recovery rather than on the ability of the canyon to assimilate impacts

• schedules for monitoring activities which, if not followed, would mean the activity (e.g., a livestock permit) would cease.

Despite its 1994 invitation to the Tracking Group to prepare an alternative, the Wallowa-Whitman NF did not present this alternative in its 1996 Draft EIS (USDA 1996). The Wallowa-Whitman NF was aware that revealing an alternative based on no-impact and least-impact options and ecosystem recovery rather than risk assessment would widen the options being considered, open up a broader public debate, and force the Forest Service to make real choices.

Since NEPA regulations require discussion of all reasonable alternatives, the Tracking Group and many members of the public challenged the Forest Service to issue a new Draft EIS that is legally sound. In February 1998 the national office of the Forest Service informed the Wallowa-Whitman NF that it could not have its Final EIS printed, because it had not presented and analyzed the Native Ecosystem Alternative.

With this decision, the Forest Service acknowledged its need to take reasonable alternatives seriously. The Wallowa-Whitman National Forest has now agreed to address the implications of the Native Ecosystem Alternative alongside its own alternatives in a new, 1999 Draft EIS. In fact, this decision prompted several local county commissioners to develop yet another alternative. Everyone involved is now more conscious of the potential for a full range of alternatives to be considered.

Organic Farmers: Alternatives to Pesticide Risk Assessment

Today in the United States, approximately 4050 commercial farmers produce food with minimal or no use of pesticides or chemical fertilizers (Dunn 1995) These farmers represent an alternative to endless risk assessments about the toxicity of pesticides (O'Brien 1988).

When sustainable agricultural operations exist, risk and benefit issues other than pesticides' toxicity to wildlife and humans come to the surface.

Reganold et al. (1987) compared the soil conditions of two adjacent wheat farms in eastern Washington, one using pesticides and the other not.

The soil was originally similar on both farms. Rain, slope, and other factors aside from agricultural methods were also similar. Both farms have been worked agriculturally since 1909. One farmer had chosen in the late 1940s and the early 1950s to begin using pesticides and fertilizers; the other had remained with organic farming, using only limited, spot pesticide spraying. To ensure soil nutrients for his crops, the organic farmer relied on crop rotation, green manure crops (i.e., crops grown in a rotation primarily because they provide nutrients to the soil rather than as commercial food crops), and native soil fertility. Compared to the pesticide-using farm, the organic farm was found to have much better soil: It had more organic content (i.e., more carbon-based materials derived from plants and animals), deeper topsoil, more polysaccharides (some of which are produced by soil microorganisms and bind soil particles together to give good structure to the soil), less soil erosion, and more soil moisture. "Our data," Reganold et al. state, "indicate that the long-term productivity of the organically farmed . . . soil is being maintained, whereas that of the conventionally farmed . . . soil is being reduced because of high rates of soil erosion." The soil scientists predicted that all the topsoil would be lost from the conventional farm in another 50 years, exposing less fertile subsoil, while the organic farm would continue to retain its topsoil. This study showed that the soil of the conventional farm could not assimilate the effects of pesticide-based farming, while the soil of the organic farm remained healthy and did not erode.

Risk assessments that focus primarily on the toxicity of a few of the ingredients in pesticide formulations do not reveal all the advantages of farming without pesticides.

Some of the clear differences between risk-assessment-based conventional agriculture and organic agriculture can be seen in the 1997 U.S. Department of Agriculture's draft rewrite (USDA 1997c) of the National Organic Standards Board's proposed national list of substances that can be used in foods federally labeled "organic." The NOSB is a group of organic industry experts that was formed in 1990 as part of the Organic Foods Production Act. The NOSB painstakingly developed a list of materials that would not be allowed under a federal organic label. Given the ultimate authority by the Organic Foods Production Act to approve the list, the Secretary of the USDA instead allowed it to be rewritten within the department. Among other things, the USDA's draft rewrite would have allowed antibiotics and parasiticides to be used in animals sold as organically grown meat. It also would have allowed irradiation, use of synthetic nutrients, application of industrial-waste sewage sludge, and use of genetically modified organisms, and it would have allowed farmers to feed meat to organically produced animals—a practice believed to be the leading cause of "mad cow" disease. The NOSB, on the other hand, would allow none of these in the production of organically produced food (Kirschenmann 1998).

The USDA proposed to provide standards for how much and which of these substances and practices could be used, and these standards would ultimately be based on USDA risk assessments of safety. In its preamble to the rule, the USDA indicated how highly soluble nitrogen, a substance previously prohibited under the organic label, would be kept within limits:

> For example, if nitrate levels in an adjacent well are found to increase over two or more crop years following application of a highly soluble mined source of nitrogen to soil . . . then the practice would have to be terminated or modified to prevent further adverse effects on water nitrate levels.

In order to eliminate the embarrassment of a meaningless federal "organic" label, the USDA proposed to allow no states or farmers to label foods according to a more restrictive, higher standard. The USDA proposed to prohibit farmers from labeling their food with such statements as "no drugs or growth hormones used" or "produced without synthetic pesticides."

A May 1997 internal USDA memorandum (USDA 1997b) revealed that one reason for its position is that the USDA did not want to appear to be implying judgment on non-organic practices or products. The memorandum indicated, for example, that if the USDA allows the organic label to exclude genetically modified organisms (GMOs), other countries will use that exclusion to claim that these organisms are not as safe as the USDA would allege: ". . . the Animal and Plant Health Inspection Service and the Foreign Agricultural Service are concerned that our trading partners will point to a USDA organic standard that excludes GMOs as evidence of the Department's concern about the safety of bioengineered commodities." The reason the USDA gave for proposing to allow pesticide residues at higher amounts than 5 percent of EPA tolerance (i.e., the NOSB limit), is that "this establishes organic as being a 'safer' food, and our program is not a food safety program."

The response of U.S. citizens was furious. The USDA received over 200,000 comments opposing this proposal. The agency announced that it was backing down, and that it would issue a revised proposal that would prohibit use of genetically engineered products, irradiation, or sewage sludge (PANNA 1998).

The existence of viable alternatives to business as usual generally raises a wide range of issues (e.g., social, value, economic, ecological, and cultural issues) that are not raised by risk assessments of a narrow range of options. This is the fundamental power of alternatives.

Citizens for Public Accountability: Alternatives to Risk-Based Permitting to Corporations

In June 1995, Hyundai Electronics America, Inc., a subsidiary of a Korean multinational corporation, announced that it would be building a $1.5 billion semiconductor factory in Eugene, Oregon, in some of the last wetlands of the Willamette Valley. The City Manager disclosed that Eugene's citizens would pay Hyundai's property taxes for three years, a total of $39 million. The facility would use 9.9 million gallons of water a day, which equals the use of 147,761 residents (personal communication, Lance Robertson, reporter for *Eugene Register-Guard*, 1998). It would use approximately 670,000 gallons of toxic chemicals a month (Robertson 1995), including ozone-depleting chlorofluorocarbons, and the extraordinarily dangerous gas arsine. If one cylinder of arsine were to be ruptured during transport, citizens living in several neighboring square blocks would be killed because arsine destroys red blood cells when it is inhaled; the only life-saving procedure is a complete blood transfusion. The combined emergency rooms of all the hospitals in most industrial communities could handle only a small number of such transfusions (La Dou 1984). Acids released into the air would fall on Nature Conservancy wetlands adjacent to the plant, and 24 acres of wetlands would be filled in for plant construction. A survey revealed that more than 135 species of plants, many of them specific to wetlands, were present on the proposed construction site. The site may be inhabited by six rare, threatened, or endangered plants (SRI/Shapiro 1995), by a fish federally listed as endangered, and by a frog, a turtle, and a butterfly that are candidates for federal listing as threatened or endangered (BPA 1995).

The semiconductor industry is non-unionized and often hires and contracts for workers for little pay. Many toxic chemicals and metals are used in the industry, including numerous solvents that are used to clean tiny semiconductor chips in "clean rooms." As a result, systemic poisoning of semiconductor workers is the highest among electronics workers (LaDou 1994). In turn, systemic poisoning is 3–4 times more likely in electronics workers than in workers in other manufacturing industries. It is higher than for workers in the chemical industry, even those in pesticide manufacture (LaDou 1994).

Upon learning of Hyundai's and the city's plans, a number of Eugene scientists, professors, lawyers, neighborhood residents, environmental activists, and citizens immediately formed an ad hoc organization, Citizens for Public Accountability, to bring the following parties to greater accountability to Eugene's workers and community and to the environment:

• the city government, which had struck the deal with Hyundai in secret
• Hyundai
• the Army Corps of Engineers, which would have to give Hyundai a permit to destroy wetlands
• the Oregon Department of Environmental Quality, which would have to give Hyundai a Clean Water Act permit
• the Lane Regional Air Pollution Authority, which would have to give Hyundai a clean air permit.

Citizens for Public Accountability proposed a Eugene-Hyundai covenant that sidesteps the details of risk-based permits for air pollution and wastewater and stormwater quality. The five elements of the covenant were as follows:

• Inform citizens of all toxic chemicals and metals Hyundai is releasing into the workplace and the community during facility operations.[5]
• Use the most worker- and community-protective technologies utilized within the semiconductor industry.
• Fund an independent community monitoring process to report on Hyundai's toxics release reporting and toxics use reduction efforts in Eugene.
• Provide secure family wages, benefits, and whistleblower protections for workers in the Eugene Hyundai facility.
• If Hyundai leaves the community within six years, pay back to the taxpayers the value of the property tax abatements the corporation received.

The first three of these provisions all go beyond current federal and state laws. In addition, the second provision would require more protection from toxic discharges than standards that are legally "acceptable" based on risk assessment. It would require recurrent analysis of technology options such as those proposed by Nicholas Ashford (see chapter 10, example 4), and installation of the most environmentally protective technologies uncovered in the technology options analyses.

Although Hyundai rejected the proposal and it was defeated in the City Council (the mayor casting a tie-breaking vote), the principles unleashed a barrage of community support and interest. A 1995 survey of Eugene citizens, for instance, which asked only about support for the first and fifth elements (i.e., right-to-know and return of tax money if the company leaves early) documented 83 percent and 90 percent support, respectively.

Though the covenant was not implemented, the City of Eugene agreed to hire a person to examine the most community-protective regulations enacted in other communities relevant to the semiconductor industry. An alternatives assessment of these regulations was presented to the City Council, which voted down proposals for more protective regulations.

Meanwhile, CPA has held labor forums, undertaken Clean Water Act and NEPA litigation, and prepared a list of long-term public-accountability issues that it intends to pursue.

One issue it has pursued successfully is a ballot initiative calling for materials accounting and asserting a right to know about toxics. In November 1996, the citizens of Eugene voted 55 percent to 45 percent to establish the first local materials-accounting law in the country (Eugene Charter Amendment 20-75). Under this law, large users of hazardous materials in Eugene must report to the public essentially all their inputs and outputs of hazardous chemicals.[6]

Materials accounting is like keeping a checkbook account of hazardous chemicals. Under the Eugene law, reported production and purchases (inputs) must equal the releases (outputs) of all hazardous chemicals a manufacturer uses, with materials accounting reporting thresholds of 50 pounds per year for hazardous substances and 5 pounds per year for extremely hazardous substances. The bill does not require the manufacturers to measure each amount of each hazardous chemical; it does require manufacturers to provide the public with their best knowledge of where their hazardous chemicals go when they use them.

Eugene manufacturers began their first year of hazardous-materials accounting on January 1, 1998, and reported their inputs and outputs for the first time on April 1, 1999. The reports are available on the World Wide Web[7] and in the public library, rather than from a government agency. A Toxics Board (three representatives of reporting industries, three advocates of a public right to know, and a seventh member recommended by the six) is appointed by the City Council to develop simple-to-use forms for reporting inputs and outputs. Manufacturers had a full year to help design the forms for reporting and to adapt their record keeping to accommodate the public's right to know.

The Montana Supreme Court: Anticipatory and Preventive Alternatives for the Blackfoot River

On October 20, 1999, the Montana Supreme Court, ruling in the case of *Montana Environmental Information Center; Clark Fork-Pend Oreille Coalition; and Women's Voice [sic] for the Earth v. Department of Environmental Quality and Seven-Up Pete Joint Venture,*[8] unanimously rejected risk assessment in favor of alternatives assessment. Its ruling was based on the Montana constitution, which affirms the right to a clean and healthful environment. In 1995, the Montana legislature had passed a law exempting water-well discharges from review under the state's Water Quality Nondegradation Law. The Montana Department of Environmental Quality had subsequently issued a permit to the Seven-Up Pete Joint Venture mining company for pumping water out of wells in an area that could yield gold. The pump water contained arsenic and reached the Blackfoot and Landers Fork Rivers, whose waters are classified as high quality.

When citizen groups protested, the DEQ said that the arsenic would dilute to "background" levels within mixing zones of nearly a mile's length, that the citizens had not demonstrated actual injury to themselves from the small increases in carcinogenic arsenic, and that they had demonstrated neither "significant" changes to the quality of water in either river nor "significant" impact on either river.

The Supreme Court did not agree that this risk- assessment approach of determining "significance" was required:

We conclude, based on the eloquent record of the Montana Constitutional Convention [of 1972] . . . that the delegates' intention was to provide language and protections which are both anticipatory and preventive. The delegates did not intend to merely prohibit that degree of environmental degradation which can be conclusively linked to ill health or physical endangerment. Our constitution does not require that dead fish float on the surface of our state's rivers and streams before its farsighted environmental protections can be invoked.

With regard to having given the Seven-Up Pete Joint Venture a waiver for its test pumps, the Court wrote that the Montana legislature had erred in passing the law allowing blanket waivers, and that "[the] government must demonstrate both a compelling state interest [in issuing a waiver] . . . and that it is the least onerous path available."

Choosing the "least onerous path available" requires analysis of alternatives.

13

The Essential Features of an Alternatives Assessment

An alternatives assessment involves consideration of the pros and cons of a decent range of options. The process should include the public whenever they might be harmed by activities considered in the assessment.

Every alternatives assessment has three basic parts: options, disadvantages of each option, and advantages of each option.

If decisions based on an alternatives assessment could result in harm to public health or wildlife, then the public must be involved in the assessment. It doesn't matter if it is an alternatives assessment done by a single public school, a single furniture factory, or a group of nations. Let us look at the three parts of alternatives assessments, and the public process.

Essential Elements of an Alternatives Assessment

Presentation of a Full Range of Options
A range of reasonable choices for behaving must be presented. The options must include those that seem to promise the least adverse impact on the environment and public health and/or those that seem to promise the greatest environmental and public-health advantages.

Presentation of Potential Adverse Effects of Each Option
The different types of adverse environmental consequences considered for the different choices must be as varied as the types of adverse environmental consequences humans are causing on Earth. The National Environmental Policy Act (NEPA) regulations, which govern the preparation of Environmental Impact Statements (EISs) by federal agencies, clearly define

adverse effects. (The NEPA regulations sometimes refer to the effects as "impacts," and other times as "effects.")

The NEPA regulations correctly note that our environment is more than simply ecological or biological functioning. This means that effects to be considered include more than illness or death. The NEPA regulation 1508.8 ("Effects") notes:

Effects includes ecological (such as the effects on natural resources and on the components, structures, and functioning of affected ecosystems), aesthetic, historic, cultural, economic, social, or cumulative.

The NEPA regulations talk of the varied scales of time and space of different effects, and of their combinations:

• direct effects, "which are caused by the action and occur at the same time and place" (40 CFR 1508.8).

• indirect effects, "which are caused by the action and are later in time or farther removed in distance, but are still reasonably foreseeable. Indirect effects may include growth-inducing effects and other effects related to induced changes in the pattern of land use, population density or growth rate, and related effects on air and water and other natural systems, including ecosystems" (40 CFR 1508.8).

• cumulative effects are effects "which [result] from the incremental impact of the action when added to other past, present, and reasonably foreseeable future actions regardless of [who] undertakes such other actions. Cumulative impacts can result from individually minor but collectively significant actions taking place over a period of time" (40 CFR 1508.7).

Using the fictional example of a proposed action to drain a 5-acre wetland in Oregon's Willamette Valley, the environmental attorney Terence Thatcher (1990), quoting Peter Steinhart, asks what should be considered when assessing the cumulative impacts of draining that wetland. He writes eloquently that the cumulative effects should include a discussion of the filling, draining, and polluting of wetlands that is occurring along the Pacific Flyway from the Arctic to Central America by myriad "minor" human activities.

While such an analysis might seem unreasonable to the developer who wants to build one factory, draining a 5-acre wetland does make a difference. This is because the damages that are occurring in the world are cumulative. Pintail ducks, for instance, are in trouble due to thousands of small activities that have eaten at their habitat throughout the Pacific Flyway. If

Box 13.1
Cumulative Effects Eloquently Illustrated

Humans joke about "being nibbled to death by ducks." Ducks may mutter more justifiably about being nibbled to death by people. Nowhere are we looking at the cumulative impacts of a lost acre this week in San Francisco Bay, another ten acres next week in Grays Harbor, a hundred in the Willamette Valley, a dozen in the Fraser River Valley, an oil development on a molting lake on Alaska's north slope. It all harkens back to our inability to see the beauty, the life, the elegance of mud and to understand that what globs and hisses at our feet is connected by invisible threads of consequence to the creatures that glide so purposefully across our skies each spring and fall. We are slowly strangling the flyway. One day, we may look out over an endless plain of concrete and asphalt and glass and find that we have drained the skies.

source: Blake and Steinhart 1987

the private landowner who wants to drain the 5-acre wetland in Oregon is not responsible for the Pacific Flyway and pintail ducks, who is?

An alternatives assessment, if it examines a broad range of options, presents the public with starkly different effects of different options. An alternative that requires no wetland draining and no pollution, for instance, might avoid all cumulative adverse effects on the Pacific Flyway, requiring mention that the option that would drain the wetland would negatively affect the Pacific Flyway. The no-drain alternative might thus (correctly) be seen as carrying enormous value.

The alternatives assessment would also consider short-term vs. long-term effects, less severe effects that affect large numbers of people and/or wildlife, severe effects that affect small numbers of people and/or wildlife, and irreversible effects.

When each option is associated with some type of adverse effects, there is no magic method of knowing which of the effects are more "acceptable." One set of people might argue for an alternative that may severely affect only one fern species, for instance, while another set of people, who know the tenuous hold the rare fern still has on Earth, may argue for another alternative that will cost a developer more money, or that might even affect several populations of three more common fern species.

When all kinds of adverse effects are being acknowledged, it becomes important to see if certain options pose markedly fewer dangers.

Presentation of Potential Beneficial Effects of Each Option

We may almost automatically think of direct, indirect, and cumulative effects as being adverse effects. Logically, however, "effects" includes beneficial effects too. As with adverse effects, the direct and indirect benefits of an option can be "ecological, aesthetic, historic, cultural, economic, social, or cumulative."

Consider, for example, social benefits of solar power. In *The Whale and the Reactor,* Langdon Winner (1986) writes of the social and cultural choices we are making when we adopt particular technologies. The choice between nuclear power and solar energy technologies, for instance, includes more than issues of waste disposal, accidents, worker and community exposure to radiation, and economics. Nuclear power involves centralized control of energy, while solar energy allows dispersed, local control of energy and energy costs. Nuclear power necessarily involves police powers and secrecy in order to protect radioactive materials against theft for the purpose of making nuclear weapons. Is there a larger environmental issue than nuclear war? One of the major environmental benefits of solar energy, therefore, is that it does not require complex police guarding.

If you were preparing a risk assessment of a nuclear power facility without considering solar power as an alternative, you theoretically could note that a nuclear power facility requires police powers and centralized control of energy. However, you are more likely to notice these adverse effects of nuclear power technology when you are also thinking about the benefits of the alternative of solar power. Decentralization of energy is often stated as a benefit of solar technologies. Consideration of this benefit of solar power reminds us of the degree to which nuclear technologies are not decentralized.

Benefits, then, can be direct, indirect, cumulative, short-term, and long-term. Unfortunately, all our efforts to allow ecosystems and organisms to retain or regain integrity are reversible. Only adverse consequences of our activities seem to have the potential for being essentially irreversible, at least short of geologic time. The willful extinction of species is irreversible, and the erosion of soil to bedrock is irreversible within the timelines we generally consider.

These, then, are the three elements in an alternatives assessment:

- a wide range of options
- consideration of the benefits of each option
- consideration of the disadvantages of each option.

If the Public Might Be Affected, the Assessment Must Be a Public Process

The Assessment Must Be Influenced by the Public

Because environment-impacting activities inevitably affect public resources, the public has a right to be involved in the assessment of alternatives to those activities.

How is meaningful involvement ensured?

We can turn once again to NEPA regulations for a model of public participation.

1. The public must be able to affect the breadth of the alternatives assessment.

Under NEPA, when a federal agency is going to prepare an environmental impact statement (EIS), the agency must notify the public and solicit ideas for the necessary scope of the environmental analysis. At this point, the public can indicate that particular alternatives to the proposal are reasonable and must therefore be described, and that particular direct, indirect, and cumulative impacts must be considered.

This process, in which the public indicates the alternatives and impacts that should be included in the EIS, is called "scoping."

A crucial step in the NEPA process is that the federal agency must respond to public scoping comments either by addressing each issue raised or by explicitly stating why a particular issue is not significant (40 *CFR* 1501.7(a)(3). In other words, the agency must do more than simply "hear" the public: It must consider the proposed alternative or impact in the EIS, or publicly explain why it is not significant enough to be included.

2. On the basis of alternatives assessment, decision makers must respond to public comments on drafts of a decision-making document.

After a federal agency has received comments from the public regarding the breadth and depth of an EIS, it then prepares a Draft EIS presenting the pros and cons of the options and usually suggesting a preferred alternative.

The agency must make the Draft EIS publicly available and give the public time to comment on its sufficiency and accuracy. The public can also comment on the proposed decision. Once again, the agency must publicly acknowledge each comment (40 *CFR* 1503.4(a)):

An agency preparing a final environmental impact statement shall assess and consider comments both individually and collectively, and shall respond by one or more of the means listed below, stating its response in the final statement. Possible responses are to:

(1) Modify alternatives, including the proposed action.
(2) Develop and evaluate alternatives not previously given serious consideration by the agency.
(3) Supplement, improve, or modify its analyses.
(4) Make factual corrections.
(5) Explain why the comments do not warrant further agency response, citing the sources, authorities, or reasons which support the agency's position and, if appropriate, indicate those circumstances which would trigger agency reappraisal or further response.

These two processes of NEPA (i.e., the early input or scoping process and the process for public commenting on a draft decision document) allow any person to suggest alternative behaviors that are reasonable and to note adverse and beneficial effects of each alternative considered. These processes yield three major social benefits.

Benefit 1: Those who are or may be adversely affected by environment-impacting activities are allowed to bring up potentially better alternatives and options.

This means that those who are managing the assessment process cannot limit the process to a few unsatisfactory choices. Those who are managing the process may not be the ones who would suffer the adverse consequences, and so they might not be thinking or caring about such consequences. They might lack crucial information about the adverse or beneficial impacts of certain options. They might want to stack the deck in favor of an option that makes their lives (or the lives of their friends or their sector of society) easier or more profitable, even though that option disadvantages people with less power or less access to the managers.

These scoping and comments processes do not depend on the numbers of commenters, or on their personal access to the managers of the assessment process. Rather, these processes are based on whether the ideas and the information brought by each commenter are valid and significant.

Benefit 2: The scoping and comments processes draw on the numerous sources of experience and alternative ways of doing things that exist in our society.

Those who are managing the assessment process might be unaware of a number of good alternatives. They might be afraid to suggest options that challenge the status quo. When assessment processes are public, we learn of options and information we had not considered before. Most communities have people with amazing skills, extensive information, and the willingness to propose innovative alternatives. Most communities have people who can suggest ways to move toward a more environmentally sound and socially just community. Many of these people will contribute to a truly public process.

Benefit 3: The public becomes more conscious of the fact that impacts on the public environment are choices about behavior, not inevitabilities.

The public becomes aware that caring better for the environment is technically and economically feasible.

What about cases in which a local community wants to accept an environmentally destructive activity and isn't interested in alternatives to that activity? For instance, a majority of the vocal public in Tooele, Utah, may have wanted to accept a number of toxic industries, including a nerve-gas-weapon incinerator (Brooke 1996).

The answer is that all affected parties need to participate in alternatives assessments. The concept of deciding locally to degrade only the "local" environment is not realistic, because local activities affect faraway people, wildlife, and ecosystems. Therefore, alternatives-assessment processes must be made accessible to the participation of affected people, who may live quite far away. In addition, local decision making is often affected by decision making at the state, regional, federal, and international levels. At each of these levels, public alternatives-assessment processes must be employed. "The public" cannot mean only those humans who live within a few miles of the decision makers.

What, then, is the limit of public involvement in each alternatives assessment? Practically, not all affected publics will have the time, the interest, or the ability to participate in all alternatives assessments. The best alternatives-assessment processes will be those that actively reach out to members of diverse publics with different perspectives, so that a full range of alternatives and a broad range of impacts and benefits of those alternatives are considered.

It is a matter of how big a circle is considered: one corporation's profits? the entire community? the surrounding state? humans and wildlife? a bioregion? a nation? the planet? Ideally, alternatives assessment embodies the bumper-sticker slogan "Think globally, act locally." But you can't think globally if you don't invite those who are globally knowledgeable to participate.

Decision Makers Must Be Accountable to a Public Process of Alternatives Assessment

A key aspect of the federal NEPA process, of course, is that the public has the ability to ensure that the assessment process is done with integrity: they can sue if it is not. Citizens can sue a federal agency if the agency has not discussed reasonable alternatives or adverse impacts that were brought to their attention during the scoping and commenting phases.[1] Without this recourse, many federal agencies would merely go through the motions of public commenting processes, and would discuss only minor adverse impacts and narrowly conceived alternatives to their preferred activity.

Alternatives to courts, at least as a first step, are also possible. If a community were entering into an alternatives-assessment process with a city council, a corporation, or some other entity, it might insist on provisions for independent arbitration in the event some in the community have reason to believe that the managers of the assessment process might evade accountability to the agreed-upon public process. Provisions for binding arbitration might be possible. Under binding arbitration, the parties to the dispute would be required to submit to the arbitrator's decision regarding public accountability. Anyone failing to abide by the arbitrator's decision could be taken to court.

Alternatives-assessment processes must provide for some form of meaningful public recourse if an unreasonably narrow range of alternatives is considered or if significant information is ignored. Otherwise, businesses and agencies would be able to get away with murder in their alternatives assessments. If a state, for instance, requires all companies to prepare or update their environmental alternatives assessments every two years, the public should have access to the schedule of such audits so that they can notify the companies of known alternatives for reducing environmental impacts. A draft of each assessment should be publicly available, again to allow the public to inform a company of reasonable alternatives it may not have considered. Documents might be made available through local libraries or electronic mail.

But how would arbitrators or courts avoid having to consider an infinite number of outlandish alternatives claimed by plaintiffs to be reasonable? A criterion by which arbitrators or courts could find a legally required alternatives assessment inadequate might be failure to discuss alternatives that are already in practice somewhere in the world for reasonably similar production or programs, alternatives that had been brought to the attention of the business or agency during the assessment process, or alternatives that would clearly have been known to the business or agency. For example, a softwood kraft pulp mill would clearly be aware, through pulp industry literature, that there are similar pulp mills that profitably produce excellent white paper without using any chlorine compounds and that therefore emit virtually no organochlorine compounds into the water, the air, the food chain, and humans. The pulp mill therefore could not refuse to include the alternative of using chlorine-free processes in an alternatives assessment of its mill.

In order to relieve each business of the work of preparing and updating alternatives assessments, trade associations of individual businesses (for instance, dry cleaners or the semiconductor industry) could prepare generic assessments, which individual facilities could then adapt to their site-specific situations.

Clearly, those who are familiar with the full range of environmental alternatives within an industry or a government sector (e.g., transportation departments) would be in demand as consultants to businesses. A profession of people skilled in various types of technology and policy options assessments would be a worthwhile profession to expand.

Some people might object that requiring public access to the alternatives-assessment processes of private businesses unduly erases the "private" from private business. There are two strong answers to this:

• Private business itself erases the meaning of "private" as soon as it affects the public. Private business is profoundly public as soon as it releases toxic chemicals into the common air; diminishes the common store of fresh water; invades the bodies of workers, citizens, and wildlife in the local community and thousands of miles away; eliminates wildlife species or their habitat; alters the hydrology of public watercourses; or fouls the world's single ocean. We have been fooled into allowing such activities to be called "private industry."

• The process of alternatives assessment does not automatically require businesses to adopt technologies that someone else judges to be the most protective of worker, a community, or the environment. Alternatives assessment simply requires businesses to talk about their options with those whom they may adversely affect. Once an alternatives assessment has been completed, other democratic processes may lead to mandated changes in technologies or permits, but that has always been the case in democratic societies. For example, an alternatives assessment showing that a semiconductor facility could operate profitably without using the highly dangerous gas arsine wouldn't mean the factory could not use arsine. However, regulatory agencies might eventually ban the use of arsine in semiconductor facilities because of the danger it poses to workers and to the public health and because alternatives assessments have revealed that there are reasonable alternatives.

Alternatives assessments must be public because all societies need to protect themselves, the environment, and future generations from destruction. National security, as has been pointed out by many people, can be threatened internally by the squandering of resources and the poisoning of life as surely as by any external military enemy.

Under a Democratic Form of Government, the Public Must Be Able to Pressure Decision Makers to Select Environmentally Sound Alternatives
The publication and dissemination of alternatives assessments obviously might lead to public pressure to implement environmentally sustainable practices. The public pressure might consist of "green labeling," consumer boycotts, laws, initiatives, or regulations. These are the processes of a democracy of informed citizens, and that is where public debates about economic feasibility would come in.

If a pulp mill, for instance, can convince its community and the government that it should continue to discharge 10 tons of toxic organochlorines into the local river every day, even though other pulp mills make excellent paper without doing so, then that would be democracy functioning.

Alternatively, communities could enact provisions requiring businesses to demonstrate significant efforts to use the most worker- and community-protective technologies available within their industry. Again, those are the possibilities of democracy.

When we realize that our lives and the lives of others are unnecessarily at stake when some people, companies, and public agencies undertake unnecessarily destructive activities, we often wake up to the potentials of democratic action.

Box 13.2
A Comparison of Alternatives Assessment and Risk Assessment

Advantages of Alternatives Assessment	Disadvantages of Risk Assessment
Focuses on a range of environmentally sustainable behaviors.	Focuses on damaging behaviors.
Focuses on available options for doing the least harm and bringing the greatest benefits to the world.	Focuses on how much of business-as-usual can be done to the world without organisms, species, or ecosystems crashing.
All sectors of the society can participate and can reinforce the efforts of other sectors.	Is highly technical and inaccessible to most members of the society.
Doesn't involve insurmountable methodological problems.	Involves insurmountable methodological problems by trying to determine safety of hazardous substances or dangerous practices and by trying to objectively rank environmental problems in relation to each other.
Involves people's creativity, innovation, and energy in supporting the world.	Involves people's defense of environmentally damaging behaviors.
Places responsibility on those who diminish, pollute, extract, and degrade to think publicly about alternative ways they can behave.	Places almost exclusive responsibility on environmental activists for thinking publicly about ways others can behave more environmentally responsibly.
Shows industries and agencies how they can avoid activities that may later cost industry and the public enormous amounts of money to mitigate (if mitigation is even possible).	Encourages industry to do activities that will cost industry and the public enormous amounts of money to mitigate (if mitigation is even possible).
Encourages industries and agencies to consider processes that are forward-looking.	Permits industry to invest in processes that likely will eventually be protested and rejected.

source: Mary O'Brien

14

A Society That Assesses Its Alternatives

In a society that replaces risk assessment with alternatives assessment, public agencies and private businesses undertake alternatives assessments, with public input.

Currently, most regulatory decisions in the United States are supposedly preceded (but in reality followed and justified) by risk assessments. Agencies and businesses reach for risk assessment as if it were a natural part of decision making. It is not. It is an optional process, and it is being used almost exclusively to figure out to what extent businesses and agencies will be permitted to damage the health of ecosystems and humans.

If we consider it a responsibility of businesses and government to instead figure out their options for treating ecosystems and public health with the greatest care, then alternatives assessment will be the decision-making tool of choice.

Let us look at how businesses and agencies should consider their responsibility to behave well toward the environment.

All public agencies, public facilities (e.g., public schools), and private facilities (e.g., factories, farms, developers, service businesses) should produce and pay for environmental benefits/impacts assessments of a range of reasonable options for their activities. The range of options should include alternatives for the following:

• no harm[1] to the environment or public health (examples of no-harm alternatives: zero discharge, closed-loop processes, use of benign chemicals, no action—e.g., no livestock grazing in a particular area of a National Forest)
• least harm to the environment or public health (examples of least-harm alternatives: significantly reduced use of toxics, major conservation of energy or water, mass transportation or an emphasis on non-motorized

transportation, tree-free paper production, return-to-producer arrangements—e.g., containers and products are returned to the producer after the product has been used or has broken)

• restoration of environmental and public health (examples of restorative alternatives: road closures in sensitive wildlife habitat, removal of unnecessary hydroelectric dams in salmon streams, phaseout of chlorine industry, replacing vacant lots with community gardens, re-establishment of natural fire regimes in forests and grasslands, actions to reach specific disease-reduction goals—e.g., 50 percent reduction of asthma incidence in an urban area, actions to promote recovery of endangered species, re-establishment of corridors connecting wildlife populations that have become isolated from others.

Example: The Eugene Public School District

In 1983, the Eugene (Oregon) Public School District established a Pesticide Policy Committee (made up of School District and City Parks Department grounds maintenance staff, an environmental group representative, and community representatives) to draw up a policy that would address the community's conflicts over schools' uses of herbicides. The Committee prepared a policy that was adopted by the School District.

One key element of the policy was that before receiving permission to use an herbicide against some vegetation on its grounds a school had to show publicly that the vegetation was truly a problem. For instance, why did dandelions have to be removed from a lawn?

Next, the school had to explain why the vegetation problem (e.g., weeds in lawns, plants in running tracks) could not be prevented (e.g., by good lawn watering and mowing techniques, or changes in track construction).

Third, if the vegetation problem could not be prevented, the school had to show whether it could be controlled manually, or with seeding, or by other non-chemical techniques.

The public can comment on and make suggestions for these alternatives assessments.

By 1996, thirteen years into this policy, what was the status of herbicide use in the Eugene School District? The following observations by Becky Riley (personal communication, 1995), a representative on the district's Landscape Maintenance Advisory Committee, are instructive:

• The Eugene School District uses very few herbicides. It installs structures that prevent the need for herbicide use, such as placing concrete strips underneath chain-link fences so that a maintenance person can mow right up to the fence. Previously, herbicides were used to kill weeds underneath the fence where they couldn't be mowed. The district has learned to place deep layers of wood chips under play areas and then roto-till weeds. They lay deeper gravel on running tracks and use a harrow to keep weeds out, and then roll the surface hard and flat again. They do not use herbicides on any lawns. Lesson: Many non-chemical alternatives work.

• The School District does not adequately continue to educate parents and staff about the herbicide policy, and yet many of these parents and staff were not around when the policy was established. Some parents, for instance, come out on weekends and themselves spray weeds in lawns. Lesson: Once a crisis is over, education needs to continue.

• Some maintenance staff would like to change the policy, but the policy includes requirements for public involvement in any decision to alter it, and the staff doesn't want to take that on, so the policy remains. Lesson: An alternatives policy benefits from clear operational procedures and requirements for public involvement.

• In spring 1995, the maintenance staff at one high school convinced the Landscape Maintenance Advisory Committee that there was no alternative to using Roundup (glyphosate) on their all-weather running track. They said the guaranty specifications of the track manufacturer required herbicide applications when weeds showed up. The staff was given permission to use Roundup. However, by the time they had completed the parent-student notification process required by the policy for herbicide applications, the weather had turned hot and the weeds died. The herbicide was never applied. Lesson: An alternatives-assessment process helps prevent knee-jerk reversion to old, hazardous, quick-fix habits.

At first glance, the process of all private and public entities preparing alternatives assessments might seem to involve endless paperwork or onerous duties, or be costly. Two responses are in order:

1. When a private business is profiting from harming the environment or the public health in any way, it is reasonable to require that business to constantly consider ways to cease harming the public environment.

When a public agency is using public money to harm the environment or the public health, it is reasonable to require the agency to periodically consider ways to cease or diminish the harm.

We currently allow industries and many public agencies to engage in behaviors that adversely affect the environment without requiring them even to consider their options for protecting it.

Should any establishment that may harm the environment or the public health be able to avoid considering ways it could cause less harm? For instance, should a dry cleaner discharge perchlorethylene into its neighborhood and community underground drinking water without publicly considering using steam (water) instead (USEPA 1998a)? Should Florida tomatoes farmers use the pesticides methyl bromide (which depletes the world's ozone layer), endosulfan (a DDT-like insecticide that leaches into the Florida Keys and is commonly found on tomatoes bought by consumers), mancozeb and maneb (which kill unwanted fungi on tomato plants or in the soil, and which cause cancer among workers, wildlife, and tomato consumers), and dozens of other pesticides that are poisonous to workers, wildlife, and tomato consumers without ever considering whether such agriculture is necessary (Davies 1996)?

If a public agency or a commercial establishment may be causing harm to the environment or the public, it has a responsibility to consider publicly its options for not causing harm.

2. Preparation of alternatives assessments need not be onerous or expensive.

Public schools in one bioregion of the country, for instance, could pool resources to prepare assessments of alternatives to use of herbicides that could then be adapted to particular circumstances in a county, a city, or an individual school.

Likewise, commodity associations (e.g., the California Strawberry Board) could pool resources of numerous farmers to examine alternatives to use of pesticides and their options for reducing agricultural runoff. Trade associations (e.g., of dry cleaners) could prepare assessments of alternatives to using toxic chemicals, and individual dry cleaning establishments could adapt those assessments to their individual situation.

What is onerous is the movement by numerous state legislatures and Congress to impose requirements for endless quantitative risk assessments before an agency can regulate private business. What is onerous is having to argue for years with industry scientists about the precise amount of harm being caused by individual substances or activities, such as dioxin, cattle grazing on public desert lands, or incineration of nerve gas weapons.

Technical options committees should be established in numerous large-scale settings to provide technical information for alternatives assessments.

The use of technical options committees by the United Nations Montreal Protocol process has provided broad international consideration of alternatives to the use of various ozone-depleting substances. Individual nations, corporations, farms, and communities are able to draw on information in the reports of these committees when planning for elimination of local uses of ozone-depleting substances.

Technical options committees could be convened by the federal government (or even internationally) regarding large national or international problems. They should be funded partially by industries involved, but also with public money so that the public participates in and has access to the entire process.

For instance, an international technical options committee might look at alternatives to generation of toxic medical wastes. This effort would be funded in part by the medical establishment and in part with taxpayer money. This process would relieve each hospital and community of researching alternatives to using so many throwaways and chlorinated plastics (e.g., PVC tubing).

Even if a technical options committee is convened only nationally, it should examine options that are in practice internationally, because advances in other nations are frequently unknown in the United States.

Example: Grafting of Annual Plants
It was interesting for many of the U.S. participants in the U.N. Methyl Bromide Technical Options Committee to learn that in Spain some annual plants such as tomatoes are being grafted, on a commercial scale, onto annual rootstock that is resistant to nematodes (a type of worm). Certain nematodes eat the roots of many tomato varieties, and then methyl bromide is often used to kill the nematodes. When nematodes kill perennial plants (e.g., grapevines), the top of a variety of the plant that produces desirable fruit is grafted onto a rootstock that resists nematodes. Numerous Spanish tomato growers have taken the unusual but commercially successful

step of grafting small annual tomato plants onto resistant tomato roots (Gómez 1993).

Technical options committees should include people with broadly different approaches to the issue being examined. Also, if a technical option can be shown to be successfully in use or otherwise clearly feasible, it must appear in the committee's report. That a given technical option is not widely used, or threatens the practices promoted by most committee members, must not keep the option out of the report.

Among the other issues that could be approached by national technical options committees are grazing of cattle on arid public lands, overuse of water in aquifers and streams, industrial use of chlorine, the mounting use of disposable products, and the extinction of salmon. Social options committees could examine options for addressing U.S. population growth, and for paying displaced workers while they go back to school to learn skills that will enable them to work at a job that provides wages, benefits, and opportunities not significantly lower than those experienced previously (Merrill 1991).

It is important that technical options committees be separated from committees that estimate the economic costs or the current political likelihood of instituting the different technical options, because it is important that we first look at what we could technically do. The Montreal Protocol process, for instance, convened both a Methyl Bromide Technical Options Committee and a Methyl Bromide Economics Committee. The latter committee examined the economics of the technical options brought forth by the former.

The work of international and national technical options committees can be refined and adapted by states or local communities.

All regulatory decision making to permit activities that could harm the environment or the public health should be preceded by publicly influenced examination of the benefits and disadvantages of a range of alternatives.

The current revolt by private industry and its legislative defenders against environmental regulatory action by governmental agencies assumes that the behavior being regulated is a given. The revolt is against regulations that feel onerous, costly, or complicated to the industry. Indeed, not only industry but also the public has an interest in avoiding the need to regulate

environmentally destructive behavior. If destructive behaviors were eliminated, complicated regulations wouldn't be needed.

It is reasonable for a government permitting agency to review publicly
whether the private interests seeking environment-impacting permits from
that agency could behave differently so that the public would not have to
run after those industries, trying to put diapers (e.g., discharge limits, scrubbers, cleanup requirements) on them.

If a society routinely used alternatives assessment to make decisions, it
would encourage private businesses to grow out of the need for regulation. Firms could accomplish this by publicly reviewing their options and
then selecting alternatives that minimized damage to humans and the
environment.

"Cleanup" and "Restoration" in a Society That Assesses Alternatives

What about the cases in which harm, even irreparable harm, has been done
to the environment and risk assessment is being used to determine how to
"clean up" a situation that can't be cleaned up—to "restore" a site that can
never be restored?

Example: Nuclear Wastes

The Hanford Nuclear Reservation, located on the Columbia River in
Washington, contains enormous amounts of radioactive and toxic chemical wastes left over from on-site production of weapons-grade plutonium
and enriched uranium in nine reactors during the period 1944–1988. Some
of the wastes eventually transformed into other radioactive elements and
chemicals. As a result, the nature and amount of chemicals and radioactive
elements present on the site aren't fully known. The location and migration
of some of the underground wastes are unknown. No one has figured out
how to completely contain some of the materials that will radiate harm for
thousands of years (USDOE 1995).

Example: Abandoned PCBs

An analogous situation exists with land and ocean contamination by
polychlorinated biphenyls (Tanabe 1988). Approximately 31 percent of
the world's production of PCBs has already been released into the global

environment. Almost 70 percent of it could eventually reach the environment. This 70 percent is in landfills and dumps, still in use in older products, or in storage. The persistence of PCBs is such that spilled, dumped, and abandoned PCBs will eventually be carried by streams from wherever they are on land to the oceans. Once in the ocean, they will be stored in fats of marine organisms and accumulated in the fats of ocean mammals and birds that eat the marine organisms. The PCBs then will be carried by these mammals and birds to the farthest reaches of the Earth and concentrated in the fats and breast milk of humans and other animals living in the most remote, least-industrialized areas on Earth. At every point along this chain, PCBs are toxic. As they concentrate in the animals, they cause more harm.

Example: Cheatgrass Invasion
So far, no one knows how to regenerate the native grasses in a sagebrush or pinyon-juniper community once the Eurasian annual grass species called cheatgrass (*Bromus tectorum*) has taken over with the assistance of livestock overgrazing (Billings 1990). Cheatgrass displaces perennial bunchgrasses, which may provide effective hiding cover for ground-nesting birds. Cheatgrass often competes with rare native forbs (broad-leaved herbaceous plants) and annuals, and yet allows invasion of other aggressive, noxious weeds, such as yellow star thistle. It increases the frequency of fires, and thus can destroy the essential shrub component of winter feeding grounds for antelope and deer (Vail 1994).

Once it appears that it will be "impossible" to clean up particular sites to pristine or "background" levels of cleanliness, or to restore native ecosystems, the polluting industries or land users generally haul in risk assessors to estimate what cleanup or restoration will be "safe" or "acceptable." They estimate how much radiation, PCBs, dioxin, or grasslands degradation, for instance, is acceptable to call "cleanup" or "restoration."

An alternatives-assessing society would address these issues of irreparable damage in two ways:

• It would prepare an alternatives assessment to analyze the technical options (initially without considering economic cost) for the greatest possible cleanup or restoration. Even if we admit we will fail to undo the damage completely, we must look at options for the most complete possible restoration. Once these options have been laid out and their different advantages and disadvantages described, one or more actions should be chosen. Realistically, the choice of action will finally be based on economic and social costs, ethics, and political will. The choice of action will be affected by the political power of those who may suffer from or who represent those who may suffer from remaining toxicity or habitat degradation. Irreparable environmental contamination and degradation, however, should also trigger a preventive response: Where are similar disasters waiting to happen?

• It would admit that a site can only be partially cleaned up or restored, and complete a survey of related production, technologies, or behaviors that could cause the same problems in the future.

Those who are going to walk away from a site that cannot be completely cleaned up or restored should be required to do two things:

• fund an independent survey of all other activities in their state or region that are likely to result in similar irreparable damage

• prepare an alternatives assessment of that industry's options for leaving no unrecallable toxic substances on any site in the future or causing no irreparable ecosystem losses.

Example: Severe Degradation of Public Lands

When agencies responsible for public lands have permitted activities that have led to landslides, noxious weed invasions, soil erosion, riparian damage, or declines in native species, the public and private users of the public lands should fund independent, publicly reviewable scientific and technical inquiries into the activities that have caused and are causing the degradation; activities that may reverse declines; and options for funding, policies, and practices that will restore the ecosystems and will hold future users of public lands accountable for continued restoration. It is frequently difficult or impossible for agencies that wish to protect past practices, past friends, and current bureaucratic arrangements to consider the full range of options that could reverse ecosystem degradation.

Example: Hazardous-Waste Sites

Any mining company that is permitted to walk away from a site or a river that has only been partially cleaned of the company's toxic mine wastes should be required to fund an independent survey of all other mining activities in the state or region that may also release toxic materials that cannot be cleaned up. It could be required also to prepare a technology alternatives assessment of its own options for never again leaving wastes that cannot be cleaned up.

These analyses, while doing nothing for the area that cannot be cleaned up, would repeatedly remind us that we are in the process of creating similar irreparable harm and would help prevent its repetition.

Many of the most passionate individual advocates of environmental and social reform are victims who cannot undo what has been done to their family or their neighborhood, or to a wildlife species, but who are determined that the suffering of other families, neighborhoods, or wildlife will halt. For instance, no one can revive a child lost at the hands of a drunk driver, but Mothers Against Drunk Driving are determined that these incidents not happen in the future. Among numerous activities (being an organization of 3,200,000 members), MADD staff and members prepare and promote state and federal legislation regarding reform of drunk driving laws, sit on law enforcement advisory boards, compile statistics, and undertake research programs (Fischer and Schwartz 1996).

The hazardous wastes dumped during the 1940s and the 1950s by the Hooker Chemicals and Plastics Corporation in the Love Canal area of New York, near Niagara Falls, have not been removed entirely from the area. Since founding the Citizens' Clearinghouse for Hazardous Wastes (now called the Center for Health, Environment and Justice) in 1983, former Love Canal resident Lois Gibbs has been helping citizens to force partial cleanups of other hazardous-waste landfills and to prevent the siting of new ones (CCHW 1992).

It is not the victims, however, who should be left to prevent the repetition of environmental or personal nightmares caused by private business and permitted by government agencies. An alternatives-assessing society routinely will require those who cause damage to examine options to prevent more damage and to help the society see where the next such irreparable damage will likely be caused if changes are not made.

Alternatives Assessment Is a Necessary First Step

"But what good does the assessment of alternatives do?" you might ask. Salmon can go extinct while we assess options for restoring their numbers.

The existence of an alternatives assessment doesn't help an inner-city child having an asthma attack right now.

It is true that assessing alternatives is not enough. One of the underlying principles of this book, however, is that one of the most essential and powerful steps to change is understanding that there are alternatives.

Private industry and protectors of environmentally damaging behaviors fear public knowledge that suffering and damage don't have to happen. They promote risk assessment because it sets loose endless divisiveness over an inappropriate question, namely "How much of a public-impacting behavior is safe or acceptable?" We have all heard variants of the following statements:

"This amount of our pesticide on your food will cause less harm than eating a peanut butter sandwich."

"You're much more at risk from smoking than from exposure to our toxic wastes."

"Who needs salmon in the Columbia River or wolves in Idaho, anyway?"

"You haven't shown that you're getting breast cancer from our PVC plastics production."

"There's a new study that contradicts these seven other studies showing harm from dioxin."

Alternatives assessment allows citizens, politicians, agency bureaucrats, and some business people to see the potential in the concept that no risk is acceptable if there are better alternatives. This allows people to move toward the next logical step: that unnecessary risks and damages are unacceptable, indefensible, and a personal or business liability. This, in turn, allows them to go to the next logical step: political action to implement change. They have an array of democratic, political processes they can consider. They might lobby, advertise, protest, boycott, write legislation, or go to court.

A father may say: "I don't care if I haven't proved that dioxin released out of your hospital incinerator caused my child in the hospital's neighborhood to be born with spina bifida. Some other hospitals are eliminating their use of materials that are thrown away and are not incinerating their other wastes."

Neighbors of a dry cleaner might say: "We don't care if we haven't shown exactly how some of us became chemically sensitive. In this town you can use water instead of perchlorethylene that might cause chemical sensitivity or other toxic effects in our neighborhood."

Citizens can use legislation to say: "I don't care if you only release one millionth of the methyl bromide in the ozone layer. You can 'fumigate' your flour mill with heat and not release any methyl bromide."

The awareness that bad behavior is totally unnecessary, that there are alternatives to behaving brutishly toward the world, is an awareness that is profoundly powerful.

Most citizens are unlikely to research and dig up alternatives to environmentally bad behavior. Many others are unable to imagine how to compel public and private entities to behave more responsibly. But many of these same citizens will find a corporation's whining about costs much less persuasive once they learn that another corporation in the same industry sector is doing just fine financially while behaving more responsibly toward the environment. They will be much more likely to stop listening to the whining corporation.

Many citizens will demand clean production, sustainable agriculture, or strong legislation once they know that attractive alternatives to dirty production and unsustainable agriculture exist. Many of these same citizens will buy a different product if they know that it means they are being more decent toward the Earth, or their family, or even themselves. Why do you think Monsanto opposes labeling to identify milk that is produced with the bovine growth hormone rbGH? It fears customer preference for milk produced without rbGH.

Alternatives assessment clearly is not enough to ensure that companies, government agencies, and consumers will change their ways of behaving. But it is hard to imagine a more powerful way of unleashing the will to change.

15
Alternatives Assessment: More Information, Fewer Pages

Because it examines the pros and cons of a range of options, alternatives assessment compiles a broader range of information. As a result, far fewer pages may be required.

A risk assessment for a hazardous-waste dump in a poor neighborhood might examine the following:

- the nature of some of the hazardous wastes that would be dumped
- some of the toxicity, persistence, and environmental fate characteristics of some of the individual hazardous wastes
- the nature of the containment of the dump (e.g., landfill liners, caps, leak detectors, back-up systems)
- estimates of air contamination by some of the toxic substances, using modeled air-dispersal patterns
- estimates of water contamination by some of the toxic substances based on models of hydrological systems (e.g., movements and gradients of water through underground strata)
- estimates of potential for spills, accidents, explosions, fire
- estimates of potential for releases of toxic substances during transportation and during the transfer of toxic wastes to the landfill
- estimates of exposure of workers and community adults and children to some of the toxic substances
- estimates of exposure of certain wildlife species to some of the toxic substances
- assessment of some health risks to workers and community adults and children posed by the hazardous-waste dump
- assessment of some health risks to some species of wildlife posed by the hazardous-waste dump.

This risk assessment might be 1000 pages long, bound in two volumes. It might include 45 formulas and cite 493 scientific studies. There might be a third volume, containing appendixes. The document would likely be incomprehensible to local residents. A review of its adequacy or accuracy might require a team made up of a landfill engineer, a hydrologist, a soil scientist, a meteorologist, a toxicologist, an epidemiologist, a wildlife biologist, and a statistician. Despite the 1000 pages of analysis, however, the report might not provide adequate information for a responsible decision about the dump.

A 50-page alternatives assessment might contain more appropriate information for decision making. This document might gather simple but accurate information, including

• evidence that all landfill liners leak (Lee and Jones 1992; Montague 1992a)

• the number and the names of the toxic chemicals that might be present in the hazardous waste

• the fact that most of those chemicals are very poorly understood, and the fact that no one knows the consequences exposure to combinations of them

• threats that have been recorded from similar dumps elsewhere

• the record of enforcement against and illegalities of the company that would be managing the dump

• specific alternative proposals to establish non-polluting, locally owned businesses in the location.

If all that information is accurate, might these 50 pages be a more appropriate basis than the 1000-page risk assessment for deciding whether to site the hazardous-waste dump in that neighborhood?

The 50-page alternatives assessment would be using a broader range of information than the 1000-page risk assessment. The alternatives assessment would include the following:

• information on alternatives to the hazardous-waste dump for providing commercial income

• information on the real-world functioning of hazardous-waste dumps

• information on what the community might expect from the hazardous-waste company in the future

• information on what is not known about the hazardous-waste chemicals.

Comparisons of Information

1. Alternatives assessment uses a broader range of information than risk assessment.

Alternatives assessment uses information about a range of alternative behaviors, evidence of benefits each alternative behavior could bring, and evidence of how each alternative behavior could cause or contribute to damages. Risk assessment, on the other hand, uses a more limited range of information, namely evidence of a few harms caused by a limited range of dangerous behaviors. Sometimes it considers only evidence of one type of harm of only one dangerous behavior (e.g., cancer risk of an herbicide sprayed on a roadside).

2. Alternatives assessment does not calculate the safety of a dangerous activity.

Risk assessors often use a lot of space when they estimate how safe some activity or substance is. They use formulas, assumptions, literature citations, and explanations. They do this even though all the hazards of that activity or substance are not well known, and even though no attempt is made to consider the effect of combining this activity or substance with other pollutants or stressful activities in the same place or inside a sensitive person or species.

Alternatives assessment doesn't have to estimate safety of any of the alternatives. It merely looks at what is known of the damages and benefits each alternative may cause.

3. Alternatives assessment considers the existence of uncertainty about any one option as a valuable piece of information.

In alternatives assessment, some alternative behaviors may be seen as undesirable simply because there is so much uncertainty about the damages they may cause. For instance, there is a lot of uncertainty about the risks of using a pesticide to defend agricultural plants. There is far less uncertainty about the risks of building soil health (e.g., through cover crops, addition of organic matter) so that plants resist pests better.

Generally, risk assessment tries to calculate safety in the midst of uncertainty regarding damages. It therefore does one of two indefensible things regarding that uncertainty:

• It may treat unproven or uncharacterized damages as "zero" in the risk calculation. For instance, a recent risk assessment by a private risk-assessment company (Ecology and Environment, Inc. 1997), prepared under contract for the State of Oregon regarding incineration of military nerve gas weapons in eastern Oregon, treated non-cancer effects of 2,3,7,8-TCDD (dioxin) that would be released from the incinerators as zero. The risk-assessment company states it did this because the U.S. Environmental Protection Agency has not established a "reference dose" (equivalent to an Acceptable Daily Intake) for dioxin in relation to non-cancer effects. However, these "non-existent" effects are acknowledged by the EPA as likely caused by smaller amounts of dioxin than those causing noticeable increases in cancer (USEPA 1994a). In other words, Ecology and Environment refused to use data from peer- and EPA-reviewed studies of non-cancer effects of dioxin for its risk assessment because the EPA hadn't supplied a single number for the risk-assessment company to plug into its assessment.

• It may "create" some number for the uncertain effects by estimating some maximum value for the damages that might be caused. The risk assessor might say "We don't know for sure, but for the purposes of this risk assessment, we'll assume that no more than 2 percent of the pesticide carbaryl will be absorbed by the skin."[1] Analogously, the risk assessor might say "We haven't observed actual cattle preference for eating this rare plant, or the effects of trampling on this plant, but we'll assume no more than 2 percent of the plant's remaining population numbers will be eliminated in this specific area of livestock grazing."

When we don't know how much damage a particular activity or substance will cause, alternatives assessment does not have to act as if the damage will be zero, and it does not have to take wild guesses at the amount of damage. It can simply say that one disadvantage of that activity or substance is that its possible adverse effects are poorly researched or understood, and therefore might be minimal, or could be serious.

4. Alternatives assessment respects potential cumulative effects by considering alternatives that would avoid or minimize such effects.

Again, alternatives assessment respects uncertainty as a form of information, a warning. For instance, a disadvantage of incinerating hazardous wastes is that we will never understand how mixtures of the numerous toxic chemicals and metals released by an incinerator will interact with people's and other animals' genetic susceptibilities. We will never understand completely how they mix with medically prescribed drugs, or other special circumstances of individual people's lives. Admitting our inability to understand the effects of mixtures of chemicals may make a non-incineration alternative more attractive. Toxics-use reduction, for instance, would avoid the "need" for more incinerators.

A risk assessor analyzing a possible hazardous-waste incinerator, on the other hand, will feel compelled either to estimate some maximum "risk" from cumulative effects or (as is usually the case) to ignore the existence of cumulative effects and pretend that each chemical being assessed is the only chemical to which people or wildlife will be exposed.

5. Alternatives assessment draws on personal knowledge.

Alternatives assessment can consider many kinds of information, because it is looking at evidence of the pros and cons of alternative options. The information considered might include a worker's experience with a particular technology, a community's experience with benefits of a particular technology, or a person's lifetime of observations regarding the difficulties encountered while trying to restore a particular ecosystem (e.g., wetlands).

Risk assessment, however, generally draws on only one limited type of scientific knowledge: evidence of limited kinds of harm related to a particular technology. Risk assessment thus tends to exclude the diverse, concrete types of knowledge and experience that many types of people have about alternative behaviors. See box 15.1 for some uses of information by alternatives assessment and risk assessment.

Box 15.1
Some Uses of Information by Alternatives Assessment and Risk Assessment

Alternatives assessment	Risk assessment
Uses a broad range of information regarding damages and benefits of a broad range of options	Uses a limited range of information about some damages caused by one or a few behaviors
Does not need to calculate the safety of a dangerous activity	Pretends to calculate what cannot be known: the safety of a dangerous activity
Treats uncertainty about potential damages of an option as a disadvantage of that option	Either ignores uncertain damages by entering them as zero damages, or arbitrarily estimates an upper numerical limit for the damages
Treats potential cumulative damages of one option as a disadvantage of that option	Either ignores potential cumulative damages by entering them as zero damages or estimates an upper numerical limit for cumulative damages
Uses personal, experience-based information when appropriate	Generally avoids using personal, experience-based information

source: Mary O'Brien

The Risk of Brinksmanship

There is an inherent danger in risk-based decision making: it might underestimate how much damage will occur, and its estimates of how much damage can be absorbed by living beings or the environment might be wrong.

Irreparable damage might occur if a risk assessment considers only one course of action and underestimates the adverse consequences of that action. For instance, a population of rare plants that a risk assessor assumed would be reduced by only 2 percent by cattle grazing might in fact be totally eliminated. A farm worker exposed to carbaryl might die because 70 percent, not 2 percent, of carbaryl was absorbed through her skin.

Alternatives assessment reminds people of choices they can make so that they don't have to run the risk of losing a plant species or of dying while picking crops.

How Much Information Is Enough?

What of the millions of pages of scientific information that have been generated in the past 15 years regarding dioxin and similar compounds? So much has been studied and estimated: Mechanisms of dioxin toxicity. Exposure routes for dioxins. Effects on mice, wood ducks, peregrine falcons, Native Americans, Inuit, trout and hundreds of other living beings. Persistence in sediments, water, soil, and humans. Production of dioxin in dioxin incinerators, cement kilns, municipal incinerators, pulp mills, the chemical industry, wood stoves, cars, and sewage treatment plants. Production of dioxin when field-burning grass crops that had been treated with herbicides. Bioconcentration of dioxin within fish. Toxic equivalencies with dioxin-like compounds. Causation of immunosuppression, cancer, endocrine disruption, P-450 enzyme induction, and behavioral effects. The list is almost endless.

Should we consider all that information in a thorough, multi-volume risk assessment of the consequences of using chlorine dioxide in a pulp mill? Would we then have an adequate risk assessment that considers more beings than humans, more kinds of damage than cancers, more dioxins than the most toxic dioxin (2,3,7,8-TCDD), more dioxin-like compounds than dioxins, more organochlorine toxic chemicals than dioxin-like organochlorine toxic chemicals, and more exposure routes than the eating of fish? How long would that document be?

Or should we read a 100-page document that summarizes the environmental, social, and economic impacts and benefits of a pulp mill that makes white paper using chlorine compounds (and therefore produces dioxin and dioxin-like compounds) relative to the impacts and benefits of a pulp mill that makes white paper without using any chlorine compounds and without using trees?

There are narrow and indefensible uses of scientific and technical information and there are expansive and socially helpful uses of scientific and technical information. The expansive and socially helpful uses don't depend on how many formulas are present, how many pages are in the document, or how few people can understand the information.

Making decisions regarding environment-impacting behaviors on the narrow basis of scientific information that is supplied by risk assessments of safety is not a defensible use of scientific information. Making decisions publicly on the broad basis of scientific and other information found in an assessment of alternative behaviors is a much wiser and defensible use of information.

III

Making the Shift to Alternatives Assessment

16
Getting Started

You can start replacing risk assessment with alternatives assessment by working to install an alternatives-assessment process in a specific situation. You can also work to replace requirements for risk assessment with requirements for alternatives assessment.

Establishing an Alternatives-Assessment Process in a Specific Situation

On any given day, thousands of risk assessments are being prepared for thousands of decisions. Alternatives assessments could be prepared instead. Any individual or group—a citizen, a CEO, a parent of a school child, a union, a citizen organization, an employee of a public agency—can work to get an alternatives-assessment process going in a specific situation.

While there is no one pattern that "fits all," the following is a generalized process for installing an alternatives-assessment process.

Step 1: Identify a specific situation in which alternatives ought to be considered.

A wetland is about to be drained. Your community is undertaking a "growth management" study. A dam is proposed for a nearby river. Pesticides are being sprayed alongside your county's roads. Atrazine is in the well that supplies your drinking water. You and some of your fellow workers are feeling sick and dizzy in your workplace. You feel that the population should not keep growing in your country. A new semiconductor facility is being given a permit to build in some of the last remaining wetlands in your community. Your region of the country once had wolves and has habitat that wolves could re-inhabit if allowed to.

Any instance of unnecessary public-health and environmental degradation would benefit from an alternatives assessment. Any regulatory or other decision-making processes that affect the environment should involve alternatives assessment.

Step 2: Determine who would be most appropriate to authorize an alternatives-assessment process for this specific situation.

Consider, for example, spraying of herbicides at your child's school. You might ask the principal to analyze their alternatives to using herbicides at the school. On the other hand, the school district might be able to do a better assessment, analyzing alternatives to using herbicides for the whole district. If several schools or school districts in your state have been successfully instituting alternatives to using herbicides, you might approach the state Board of Education to urge a statewide alternatives assessment.

For a second example, consider incineration of chemical weapons. Perhaps your local group could work to require an assessment of technological alternatives available for dismantling chemical weapons at a local Army depot where construction and use of an incinerator are being proposed. This type of assessment has been urged by the Confederated Tribes of the Umatilla Indian Reservation, in eastern Oregon, in relation to the proposed chemical weapons incinerator at the Umatilla Army Depot. On the other hand, a nationwide organization of such groups might urge such an assessment by the Army for treatment of all of its chemical weapons. In the United States, this step is being urged by the Chemical Weapons Working Group. Alternatively, the Chemical Weapons Working Group might urge that the National Research Council convene a committee to examine these alternatives. The National Research Council, an organization of volunteer scientists, engineers, and other professionals, establishes special committees to look at particular issues that have large scientific and social implications. A committee of the National Research Council is in fact currently examining alternatives to incinerating chemical weapons.

Your choice of whom to pressure to authorize the alternatives assessment will be affected by questions such as these:

• Who should be doing the alternatives assessment? For instance, if a factory is seeking a permit to pollute the air of your community, you and your group may feel that it is the responsibility of the factory to hire an independent auditor to assess the factory's potential for pollution reduction.

• Who would undertake the most thorough and candid alternatives assessment?

• Who would use the best public process during the alternatives assessment?

• Who can draw on the expertise needed to do an alternatives assessment?

• Who is willing to undertake an alternatives-assessment process?

• Whose alternatives-assessment process would be taken most seriously by the public and/or by decision makers?

Sometimes getting an initial, simple alternatives assessment done will help highlight the need for a more extensive alternatives assessment. Your group may even try to develop an alternatives assessment in order to show the utility of a more public, comprehensive process. However, you want to avoid, if possible, providing the only alternatives assessment, because it might be ignored.

Step 3: Convince the appropriate body to undertake the alternatives assessment.

Once you have identified whom you want to undertake the alternatives assessment, you have to organize to get them to agree to do it. Depending on the situation, you might do one of the following:

• Explain exactly what you mean by an alternatives-assessment process.

• Describe how the alternatives assessment could be undertaken without consuming excessive time or expense.

• Emphasize any provisions in relevant local or state laws or policies, or in federal laws that encourage their consideration of alternatives in some form. Even though terms such as "alternatives assessment" or "technology options analysis" are not likely to be present in the laws or policies, anything that requires public input or assessment of impacts can logically be interpreted to require consideration of options. Any provisions for cost-benefit analysis, for instance, could be used to argue for consideration of social and environmental costs and benefits as well as economic costs and benefits. Many large U.S. corporations subscribe to what they call "continuous improvement" in their environmental performance. They like to keep the improvements on their terms, however, and most don't make their environmental audits public (Roht-Arriaza 1995). You can capitalize on their lip service to continuous improvement to encourage them to undertake continuous improvement in the public light. After all, they affect the public with their toxic chemical releases and potential chemical accidents.

• List all possible advantages to both the community and the alternatives-assessment body of undertaking the process.

• Describe examples of attractive or innovative alternatives that might be considered in the process.

• Indicate other bodies that have successfully undertaken alternatives assessments on similar issues with useful results.

• Adamantly oppose a risk-assessment approach, because it will not address issues of concern to the public. Insist that making a decision based only on risk assessment is like sitting on a one-legged chair. It is not a sufficient base for making decisions.

• Reveal the lack of candor or accuracy in statements by those who oppose an alternatives assessment.

Step 4: Participate in shaping the alternatives-assessment process.

Make sure that the following things are happening:

• Appropriate questions are being asked. For instance, alternatives assessment for making paper with less environmental impact might ask any of the following questions: "What are the alternatives for making ISO 90 (i.e., bright white) paper?" (With this question, probably only chlorine-based processes will be considered.) "What are the commercially available alternatives for making paper that consumers have shown a willingness to buy?" (With this question, only established processes will be considered.) "What are the alternatives for making paper without using trees and without using chlorine compounds?" (With this question, all reasonable alternatives, including those nearing commercial viability, can be considered.) The scope of the questions asked will influence the range of alternatives that are considered.

• An appropriate range of alternatives is being considered.

• Advantages and disadvantages of each alternative are being considered. These must include biological, social, economic, and political advantages and disadvantages for both the short term and the long term.

• The public is being encouraged to participate, and the language being used is understandable by community people.

• Good options are actually being presented during the process. Communicate with those who know about alternative technologies or other options being considered. Even if they won't participate in the process directly, encourage them to provide the information you need in order to present the option they know about. Show them materials you prepare so they can comment and make suggestions. Involve them as much as you can.

• Drafts of the alternatives assessment are being prepared for public comment.

• Claims that are made in the assessment regarding impacts of the alternatives are being backed up with evidence that is available to the public for review and comment.

Step 5: Keep in mind that, in addition to wanting an alternatives assessment, you want to contribute to the wider social process of replacing risk assessment with alternatives assessment.

In order to contribute to the wider social process, consider doing the following:

• Communicate with other groups about the process you are undertaking and the successes you are having.

• Communicate your conviction that the alternatives-assessment process that is being undertaken in your situation is what should be happening systematically throughout your community, state, and nation.

• Use the example of your alternatives-assessment process to challenge risk-assessment processes for other situations.

Building Frameworks That Include Alternatives Assessments

Ultimately, we need to design and implement laws, regulations, and public and private policies that encourage and require public assessment of alternatives. For years, the business community, the "wise use" movement, and the political right have been working to pass federal and state laws that require extensive risk assessments prior to any regulatory action. We can build on this movement by saying "It's good to look at risks, but then isn't it even better to look at all kinds of important disadvantages (risks) of various options, and their advantages (benefits) as well?" Although legislators, government agency people, judges, and some of the public have been going along with the belief that risk assessment is a reasonable way to make decisions, it would be difficult for them to deny that if it makes sense to look at certain risks, it makes sense to look at a broad range of risks of a number of options and at the advantages of those options as well.

While you are working to install alternatives assessment in place of risk assessment, you will probably run into resistance. Depending on the situation, you may be moving against any or all of the following:

• habitual, risk-based decision-making processes
• private profit or greed, which will lead certain people and businesses to want things like profits and privileges to remain unchanged
• lack of imagination
• timidity
• bureaucracy
• racism
• entrenched political power, which wants to avoid public consideration of alternatives
• business as usual.

In the following pages, I provide some ideas that may help you remind yourself why you are pushing for alternatives assessment, and help you keep going with energy and determination.

Keeping Going

Remind yourself frequently of the following:

• You have a right to good health while living in your environment (air, water, soil, food, sunlight). Others, such as developing embryos, people of little material affluence, wildlife, and trees, have equivalent needs and rights to health and habitat as well.

• Despite the claims of certain corporations and individuals to the contrary, no one has the right to take for themselves at the expense of the whole community.

• No imposed risk is acceptable if alternatives that do not impose risk are available.

• Scientists are not being scientific when they claim that a certain hazardous activity or substance is safe. At the very least, they aren't considering the cumulative effects of that hazardous activity or substance combined with the effects of other hazards being experienced. (See chapter 4.)

• Your aim is to help people behave better toward the Earth and each other. Processes (such as risk assessment of narrow options) that end up permitting unnecessary hazardous activities as safe are not acceptable.

• The aura of professionalism you might gain by involving yourself in arcane, quantitative risk assessments comes at the expense of the Earth and its inhabitants. Becoming deeply involved with unrealistic numbers and manipulated processes around a pinched range of options is not professional in any meaningful or worthwhile sense.

• You may be getting sucked into risk assessments of single options or an inadequate range of good options. Risk assessment is now so pervasive that it can take effort even to be aware that you are participating in what is essentially a risk-assessment approach. The essential test of whether a process is an alternative assessment is the question "Are all reasonable options being publicly considered for their benefits and drawbacks?" If the answer is No, change the process, or at the very least refuse to put your personal time and energy into it.

Draw on the work of others:

• Keep in touch with people who realize that risk assessment is a diversion and who advocate alternative processes. Read *Rachel's Environment & Health Weekly,*[1] for instance. Read Greenpeace project materials. Work with people who are being adamant about saying "No" to some project or substance, or people who are working to replace some hazardous activity with an excellent, life-protecting alternative.

• Keep in touch with at least one friend who will know and tell you when you are straying into risk-assessment thought patterns or processes.

• Collect and store examples of where alternatives assessment has been employed. This can be tricky, because the situations in which the pros and cons of a good range of options were considered will probably not have been called an "alternatives assessment." What it is called doesn't matter; what matters is that the essential features of alternatives assessment are present. (See chapter 13.)

• Collect and store examples of where damages or risks have turned out to be worse than had been estimated by risk assessors and industry. These examples don't have to be in your area of environmental work at all. For instance, the fact that estimates of the levels of radiation considered "safe" have steadily plummeted over the years (see figure 3.1) is a classic example that can be used even if you are working on butterfly conservation or saving wetlands. The Intel and IBM risk assessments regarding frequency of mistakes that a user would encounter using the Pentium chip (chapter 2, case 4) is a classic in another way. Another example involves the 1986 explosion of the space shuttle *Challenger*. The National Aeronautic Space Administration had pressured the manufacturer of the booster rockets, Morton Thiokol, to give the go-ahead for the launch of *Challenger* because the data were "inconclusive" as to whether the O-rings would fail to seal in hot gases during cold weather. Shortly after being launched in 36°F weather, *Challenger* was engulfed in flames (Hoversten 1996). Requiring "conclusive data" before caution would be invoked resulted in the utter failure of the launch, the loss of seven lives, and a lifetime of regret for those who did not insist on caution. Morton Thiokol had simply urged that the launch be delayed until afternoon, when the weather would be warmer; it had not insisted on caution (Hoversten et al. 1996).

• Collect and store examples of where positive change in behaviors toward the environment have occurred. Almost certainly, positive changes in behaviors toward the environment have involved some sort of alternatives assessment. A rancher may have changed how she treated habitat for wildlife; a company may have eliminated its use of a toxic chemical; a community may have rejected the inevitability of "growth." There is a great deal of power in these positive stories. They remind people that others have managed to change their behavior in ways that cooperate with the environment. These affirmations weaken people's stance that the change you are seeking could not occur.

• Collect and store examples of good anti-risk-assessment and pro-alternatives-assessment speeches, slogans, and graphics. For instance, for a public rally in the spring of 1995 protesting the 104th Congress's Contract with America legislative damage, Greenpeace prepared a T-shirt with the slogan "Are you 1 in 1000?" To publicize its opposition to the Dow Chemical Company's profitable manufacture of chlorine, Greenpeace prepared a large banner reading "Dow Shall Not Kill." Look to these examples for good anecdotes, stories, persuasive evidence, creative inspiration, and hope.

Communicate with energy:

• Bring as much media attention as possible to any refusal or failure by industry, agencies, or other decision makers to consider a decent range of options. Point out that their refusal is not reasonable in view of what might be at stake, and that they may be afraid of certain options. If you criticize them for being afraid to even discuss options, they may be less likely to repeat that refusal in the future.

• Clarify a risk-assessment situation or an alternatives-assessment proposal with analogies, phrases, slogans. Remember, no imposed risk is acceptable if the risk isn't necessary. Several analogies are presented in this book, including the woman by the river (chapter 1), the risk assessment of pollution balls (chapter 1) the tortured spy (chapter 2), and the "safe" Bloody Marys (chapter 4). As an exercise, say to yourself "What they're doing here is like . . ." and then complete the sentence with an analogy. It helps if you can think of an analogy everyone will understand, or around which there is strong consensus in the community, or that will strike home with one or more key decision makers. Just make sure that the analogy

I WANT YOU
TO ADOPT A CHEMICAL

Figure 16.1
Source: brochure for National Cancer Volunteer Registry, Agricultural Resources Center, Carrboro, North Carolina. Reprinted courtesy of Agricultural Resources Center. Original artwork by Glenn Dodenhoff. Copyright Agricultural Resources Center.

holds throughout, and that it is reasonable and not offensive. Try it out on a few friends or co-workers. Try it out on someone who is representative of your audience. See if your devil's advocate can think of ways the analogy doesn't hold.

• Appeal to the values and emotions of people. (See figure 16.1.) This is precisely what promoters of risk assessment try to avoid so that unnecessary death, suffering, and destruction can appear "rational," abstract, and justified by numbers. Rachel Carson's book *Silent Spring* is perhaps the most superb example of joining science, common sense, emotion, and values. Read or reread it so that you will be reminded of the legitimacy and power of emotion linked to information. Emotion doesn't mean irrationality; it means that decisions about how we as communities, governments, and businesses will treat the Earth and each other must not be justified by numbers manipulated on calculators. Numbers do not and cannot address what is whole, ineffable, priceless, or just. They cannot address

Box 16.1
Risk Assessments Have Real Consequences

Many of the victims of modern chemistry are literally fighting for their lives, and the lives of their children. They are real human beings with real stories to tell. They are not "hysterical housewives." They are mothers of children with Hodgkin's disease, or leukemia or chronic bronchitis, or with chemical sensitivities so bad they need to carry an oxygen tank to the supermarket. Or they are people with emphysema or eczema or any of a hundred other disabilities brought on by the chemicals spewed into our homes, or workplaces, our schools by the "better living through chemistry" mentality. Some are people who have a neighbor or a friend whose life has been disrupted, in some cases, destroyed, by toxic exposure. Their message is urgent and compelling and simple: we want justice, an end to the pain, the suffering, the carelessness and cruelty of the users and dumpers of toxic chemicals.

source: Montague 1988

hope, discouragement, respect, anger, aesthetics, sense of place, longing, or the waste of unnecessary illness or dying young. The numbers of risk assessment are abstractions of degradation; alternatives assessment incorporates the concreteness of caring.

• Communicate your intention that alternatives assessment be done. It is risk-assessment-based decision making that is bizarre and alien to common experience, not assessments of alternatives prior to decision making. Have confidence.

17

Barriers to Alternatives Assessment

Alternatives assessment faces the same types of barriers that exist on the way to every social process of caring more for others.

It sounds so reasonable to say "Publicly debate the pros and cons of good alternatives before making decisions that affect the environment and public health." It also sounds reasonable to propose that women vote, that workers have effective workplace organizations, that water be clean, and that wildlife species be allowed to survive. Trying to get there, however, is consistently met with fierce resistance.

Unfortunately, alternatives assessment is resisted for many reasons. These barriers to alternatives assessment must be taken into account in any strategy to institute alternatives assessment in any setting.

Barrier 1: Risk assessment has become a standard way of thinking in the United States, so the idea of structuring decision making around alternatives assessment often does not occur to people.

During the 1970s, risk assessment became "the way" to justify permits for technological hazards. The Environmental Protection Agency, the Food and Drug Administration, and the Occupational Safety and Health Administration in particular were effective promoters of risk assessment. In the following passage, Peter Montague of the Environmental Research Foundation describes how, in about two decades, public agencies, large environmental organizations, and private industry all constructed or joined the risk-assessment bandwagon, each for different reasons (Montague 1995):

Initially, risk assessment was devised as a rational response to technological hazards. Government agencies thought cost-benefit analysis and risk assessment offered a science-based and information-based approach to public policy, as distinct from the older form of party- and machine- and pressure-politics as a way of formulating public policy.

The modern environmental community, which came into existence in the late 1960s, for the most part adopted this rationale without question. Large environmental organizations with staffs of scientists began to operate within the "risk assessment" arena.

It was not long before the purveyors of hazardous technologies realized that risk assessment could be used to (a) exclude the public from decision making because for the most part the public doesn't think numerically; (b) make any project or technology, no matter how outrageous, appear to be "acceptable" because the assumptions going into a risk assessment could be readily manipulated, yet the final outcome always had a satisfyingly scientific appearance; and (c) allow the polluters to lay claim to "good science" and to accuse their adversaries of being emotional and scientifically illiterate.

By 1995, industry was proposing that all aspects of environmental regulation be based on quantitative risk assessment. In the first session of the 104th Congress, the Republican majorities proposed that quantitative risk assessment be required in the process of making all federal environmental regulations.[1] Meanwhile, conservative majorities in many state legislatures were proposing the same requirement for developing all state environmental regulations. By proposing extraordinarily complex and detailed quantitative risk assessments for nearly all proposed regulations, and by requiring public agencies to essentially gain approval for each risk assessment from the industries they regulate, these conservative majorities were seeking, in essence, to halt the creation of environmental regulations.

But these blatant efforts to require all agencies (whose budgets are at the same time being cut) to prepare complex, time-consuming risk assessments are only an absurd extension of what has already become a reality in most regulatory agencies: when a decision is to be made, a formal (or informal) risk assessment is prepared and debated, and the eventual decision is then backed up by that risk assessment.

The process has become so commonplace as to seem almost natural. The most common question I am asked when talking about the need to do alternatives assessments is "But don't we need to do risk assessment of the alternatives?" My answer is "Yes. Risk assessment is part of looking at the disadvantages of each alternative."

However, it is interesting that most people, when faced with a risk assessment, don't also ask "But don't we need to look at alternatives to these risks?" In other words, most people don't imagine not doing a risk assessment. When these people are handed a risk assessment, they don't generally notice that alternatives to risk aren't being considered. It doesn't occur to them to object to the limited discussion of risks they may be required to bear.

Many people have become so used to seeing risk assessments for bad options that they forget that any other approach is possible.

Barrier 2: Entrenched interests and behavior patterns are threatened by alternatives assessment, because it makes their environmentally bad behavior seem a social problem.

As one political scientist (Stone 1989) notes, "there is an old saw in political science that difficult conditions become problems only when people come to see them as amenable to human action." In other words, once we know that there is a bridge across an icy river, we are less likely to obey someone who is telling us to wade across. However, the ones who want us to wade across (e.g., people who make their living off a warm café on the other side of the river) don't want us to know that there is a bridge farther upstream.

When an alternatives assessment describes reasonable alternatives that would bring relief from degradation of environment and health, people are more likely to try halting the degrading activity.

For a variety of reasons, most people are not skilled at imagining alternatives to bad technologies, or to overconsumption, or even to population growth. But some people are good at imagining and proposing feasible, attractive alternatives. And a lot of people are good at recognizing good ideas when someone else brings them up. If we provide opportunities for the people with imagination and experience to lay out attractive alternatives in front of the public, then there are a lot of people who will recognize good, reasonable, fair alternatives.

This alternatives-assessment process spells trouble for those who don't want change. Therefore, people defending the status quo try to make sure that the process of laying out alternatives doesn't take place, or at least doesn't take place in public. This is why, from the standpoint of entrenched interests, numerous questions are better left unasked: Why should we make paper out of trees when we could make it out of agricultural wastes (Moran 1989)? Why should we deposit our bodily wastes in water rather than returning them to the soil via biological toilets (see chapter 6)? Why should we add another highway bypass in our city when a light-rail system would obviate the need for more cars and more roads for cars?

Barrier 3: Sometimes the only reasonable alternatives are more "difficult" to use or more expensive to establish.

Those who want to protect established behaviors, no matter how environmentally "costly" in the long run, will bring out roadblocks to prevent discussing an alternative, such as that it involves special training or skills, new kinds of expenses, or costly initial investments.

For example, it requires more skill to farm organically than to simply spray pesticides according to the calendar or on the advice of a local agricultural extension office, or to apply chemical fertilizers according to a label. Ecologically sound farming involves knowing the soil conditions of one's farm intimately and knowing the differences between "problem" insects, mites, nematodes, or weeds and "beneficial" insects, mites, nematodes, or "weeds." It requires close observation of how one's crop plants are doing in different parts of the field. It involves trial and error, because the farmer can't use recipes from the chemical companies, and many state university agricultural sciences department don't research or know about nonchemical farming techniques. The claim is then falsely made that ecologically sound farming is not feasible.

Enabling a facility to deal with its harmful by-products through closed-loop or zero-discharge practices might involve major initial investments. Corporations have a legal obligation to their shareholders to provide a high return on their investment. Making the environment absorb, for free, many of the true "costs" of the corporation generally allows the shareholders to increase their personal, private "return."

Certain corporate behaviors that would provide relief to the environment, on the other hand, might reduce, at least temporarily, the shareholders' financial return. The current legal fiction of a corporation generally prevents wide-ranging alternatives assessment and guarantees externalization of costs onto wildlife, communities, the environment, and workers (Montague

1994b). Corporations are unlikely to engage in wide-ranging alternatives assessment, because increasing their short-term profits requires externalizing their costs onto wildlife, communities, the environment, and workers.

Society, of course, can decide to provide a different bottom line: that the corporations will not be allowed to develop a profit for their shareholders unless they drastically reduce or eliminate externalities,[2] or unless the corporations inform the public (i.e., the market) completely about what each corporation is doing to the environment. Society could require, for instance, that chemical companies reveal all the chemicals that are present in their pesticide formulations. It could require each business to report the names and amounts of all chemicals it releases into the environment. Society (or even a single community) could also require that businesses publicly report what they could be doing for the environment (i.e., through a public alternatives assessment or technological options analysis).

Barrier 4: Sometimes risk assessment seems to come out on the side of the environment, making public-interest groups reluctant to attack the risk-assessment concept.

Sometimes a short-term benefit to the public interest is gained on the basis of a risk assessment, reinforcing the public's acceptance of risk assessment as a tool. For example, benefit was gained when the Natural Resources Defense Council worked with CBS Television to produce a *60 Minutes* program on the danger of cancer posed to children by Alar (daminozide) and other pesticides in their food. The pesticide Alar is used to keep apples on trees longer for simultaneous picking and to retain shelf life. In experimental studies, Alar has been shown to be a potent carcinogen (USEPA 1984a). Apple juice and applesauce prepared from Alar-treated apples contain Alar.

The NRDC has raised concerns for years regarding the toxic effects of pesticides. One of these concerns is that small children may suffer especially high damage from certain pesticides that are present in the foods they eat. Children's bodies may be particularly vulnerable to certain toxic pesticides. Also, children eat relatively large amounts of certain foods, such as apple juice and applesauce, that contain EPA-permitted residues of certain pesticides, such as Alar. If the first steps of cancer are caused early in a person's life, the full-blown cancer may appear at quite an early age.

In 1989, the NRDC prepared a report on these types of problems with pesticides in children's foods, and illustrated the problem with the story of Alar (Sewell 1989). Using information on the toxicity of Alar to prepare a risk assessment, the NRDC had demonstrated that Alar is extremely toxic. CBS aired this rather dramatic story of the risks to children from Alar, the willingness of the EPA to permit a highly carcinogenic pesticide on children's food, and the lack of necessity to use this pesticide. Viewers were shocked. In the ensuing uproar, the apple industry voluntarily dropped its use of Alar, and more of the public became aware of dangers posed by pesticides in food. At the same time, however, the NRDC had, at least inadvertently, presented risk assessment as a useful tool.

But what about the 21,500 other pesticide formulations for which dramatic risk assessments are either not available or not likely to be conducted (USEPA 1994d)? Should most pesticides stay on the market because they cause only a few cancers, a little bit of immune suppression, undramatic birth defects, or a few miscarriages compared to pesticides like Alar?

The reliance by the NRDC on risk assessment to show that Alar was highly dangerous implied that risk assessment is accurate. In fact, the NRDC was arguing that their risk assessment of Alar was accurate. They were arguing that the public should put trust in their risk assessment of Alar. This implies that risk assessment, at least if done "right," is worthy of trust. The public could logically conclude, then, for instance, that a risk assessment that purports to show that the herbicide Roundup (glyphosate) is "not very toxic" indeed legitimizes Roundup.

In reality, the NRDC is acutely aware that the entire risk-assessment process is riddled with assumptions, manipulation of data, incomplete toxicity testing, inability to determine risks of mixtures of toxic substances, and industry risk assessors ("tobacco scientists") who will always show that any toxic chemical really isn't toxic. The NRDC made a strategic decision to rely on risk assessment when it served their purposes of showing the dangers of certain pesticides to young children. In the process, they may have done much to bolster public trust in risk assessment, which is generally used to defend pesticides in children's food.

Meanwhile, the ban on Alar may have resulted in farm workers being more exposed to other toxic chemicals, including insecticides, used in an effort to prevent mite damage that some farmers feel causes early drop of apples on trees not treated with Alar (Rosenberg 1995). Production of daminozide, the active ingredient of Alar, may not have decreased follow- ing the ban: greenhouse workers and workers in manufacturing plants con- tinue to be exposed, and some unknown amount of daminozide is exported (Rosenberg 1995). A risk assessment of agricultural, trade, and production practices following the ban has not been prepared.

In defense of risk assessments, some people will point to the fact that the most recent (1994) EPA risk assessment of dioxin showed that it is even more dangerous than had been previously thought. However, assessments of various chlorinated products derived from chlorophenols (e.g., pen- tachlorophenol) have shown them to be dangerously toxic since the mid 1930s (Whiteside 1979), and EPA dioxin assessments have been nearly con- tinuous since 1979 (Webster and Commoner 1994).

Intermittently from 1970 to 1983, and intensively from 1978 to 1983, the EPA examined the risks of the herbicide, 2,4,5-T. Dioxin was the focus of this process because 2,4,5-T is contaminated with dioxin (Whiteside 1979). In 1979, EPA assessed the risks of dioxin when a waste-to-energy incinerator in Hempstead, New York, was making workers in a Federal Aviation Administration building downwind of the incinerator sick. From 1980 to 1988, EPA went through an exhausting (and politically compro- mised) regulations process regarding pentachlorophenol wastes, which are contaminated with dioxins (Van Strum and Merrell 1989). The current reassessment of dioxin risk began in 1991, after some industry scientists had convinced EPA that the risks of dioxin may not have been as high as had been thought (Bailey 1992). Ironically, the 1994 risk-assessment process had shown the effects of dioxin to be more severe than had been previously thought (USEPA 1994c; Roberts 1991). The reassessment is con- tinuing as of 1999.

Many hope that this risk-assessment process will finally show dioxin to be so toxic that we can finally eliminate dioxin production. And these hope- ful people may be right or wrong. But what else are we learning in this process?

Are we learning to believe that the risk-assessment process is the appro- priate process to rely upon before regulating a toxic chemical? Countless hours have been spent by EPA scientists as well as non-EPA scientists, indus- try risk analysts, volunteer advisory committee members, and citizens in dioxin risk assessments in the United States alone during the past two decades.

Do we think that we now understand the toxic risks of chemicals in our society? There are tens of thousands of chemicals in use by industry about whose toxicity essentially nothing is known (USEPA 1998b).

Or in fact are polluters getting us to fight for years over the exact details of risks of each toxic chemical while they head to the bank to deposit their profits gained while producing and discharging each chemical into the world and into us?

Barrier 5: Risk assessment looks "scientific" and "rational"; alternatives assessment may look "messy."

Alternatives assessment involves multiple options. When it is done publicly, we often see the gulfs between what different groups (e.g., a factory and wetlands defenders, or a timber company and local sustainable foresters) understand and value. We often see more clearly who suffers and who benefits from each of the alternatives being considered. Alternatives assessment involves the public and their values and emotions as well as their numbers and scientific information.

Risk assessment can seem more attractive than alternatives assessment, because risk assessment can look "scientific" and "rational" while alternatives assessment appears more socially painful.

The hope for risk assessment being scientific and stress-free, however, is illusory. Birds keep dying. People keep getting sick. Articles keep being published that show the political and profit-driven biases in a given risk assessment. Citizens keep protesting "risk-based" decisions to permit some hazardous activity.

In addition, it is naive to think that decisions regarding hazardous activities are being based primarily on the risk assessments that justify them. The decisions may be based in part on risk assessment, but they are in large part made on the basis of major factors not related to risk assessment: political power, corporate power, ideologies, misinformation, values, business as usual.

Meanwhile, the outcome of a risk assessment varies depending on the supervisor of the risk assessor or the entity that is paying for the risk assessment.

Risk assessment, with its weighty cloak of numbers and technical terms, tends to hide numerous social elements driving decision making. In reality, risk-based decision making is just as socially messy as decision making based on alternatives assessment. It's just that alternatives assessment helps make visible the non-scientific elements that are always behind risk-influenced decisions regarding who will be allowed to do what to the environment.

Barrier 6: Alternatives assessment can seem too open-ended.

If all reasonable alternatives for an activity must be considered, what end is there to the possibilities? This is an interesting question, because the range of alternatives will be limited or enlarged depending on the stated scope of the decision. If the ostensible scope of a decision is to address "waste," for instance, then at least one alternative would properly address reducing the generation of that waste. If, on the other hand, the scope of the decision is limited to whether incineration is the "cheapest" method of waste disposal, then the alternatives have been critically narrowed. (If the scope is only whether this incinerator meets legal standards, then it is not even an alternatives assessment.)

What about the range of alternatives that should be assessed by an existing commercial facility? For example, should a company that makes chlorine publicly consider an alternative technology that would eliminate the need for the company altogether? That's another interesting question, and it points to why alternatives assessments ought to be used at multiple levels in our society. The chemical company will most likely not consider alternatives to existing. Its alternatives assessment may address limiting its discharges to the environment, reducing the likelihood of a chemical accident at the plant, or perhaps reusing particular intermediate chemicals. At the level of the nation, however, the question of replacing the use of chlorine is legitimate, and an alternatives assessment regarding which sectors of the chlorine industry can most easily be replaced and which are difficult to replace may ultimately result in decisions that lead to shutting down particular facilities. Likewise, alternatives assessments at the national and state levels can address options for helping workers become retrained so they can obtain other long-term, well-paying jobs if they are displaced by the shutdown of their industry.

In reality, alternatives assessment is more open-ended than risk assessment. More alternatives are considered, but the number of feasible proposed alternatives will not be infinite. If we are afraid of alternatives assessment because we would have too many options, then we are ultimately afraid of making responsible decisions. A democracy, by the same logic, is more open-ended than a dictatorship.

We must remember that we lose a lot when we narrow our choices. At the very least, we lose our sense of being citizens in a democracy; at the worst, we lose our sense of being responsible decision makers about our own lives and about the future of life on Earth.

Barrier 7: Since alternatives assessment is open-ended, it may fail to consider legitimate, reasonable alternatives.

As a flip side to the seemingly endless number of legitimate alternatives, there is the potential that the alternatives considered will not cover a broad enough range. For any number of reasons, certain interest groups, businesses, government departments, or individuals will want to keep certain alternatives off the table. By simply pronouncing a particular alternative "unreasonable," some decision makers, economic interests, or even courts can avoid public consideration of reasonable alternatives.

Another reason legitimate, reasonable alternatives may not be considered is the lack of knowledge of many people regarding alternatives. For instance, because many public agencies are currently organized around risk-based permitting of activities, what does the permit writer do if he or she knows only one technology? Though he may be able to construct a risk assessment of a specific facility, such as an incinerator, a permit writer may not be equipped to oversee an alternatives assessment, because he may know nothing about feasible alternatives to incineration, or alternatives to generation of waste.

This problem is seen within the EPA's Office of Pesticide Programs, which basically registers pesticides. The Office of Pesticide Programs staff knows about registration and re-registration of pesticides, animal experiments on the toxicity of pesticides, and record keeping about pesticides. However, most members of the staff do not know about *alternatives* to using pesticides, such as bio-intensive agriculture, integrated pest management, organic agriculture, or sustainable agriculture. Therefore, most of the staff of the Office of Pesticide Programs would have trouble helping with an assessment of alternatives to pesticides.

In 1995, the EPA proudly announced that the Office of Pesticide Programs had formed a "pilot multidisciplinary division" responsible for (1) registration and re-registration of biological pesticides and (2) integrated pest management programs (USEPA 1995). The group was to be responsible for developing activities to prevent pesticide pollution, overseeing the development and use of "safer" pesticides, and accelerating the registration of new biological pesticides. According to an EPA spokesperson, there are perhaps 50–60 employees in this division. Meanwhile, the other 800–900 employees[3] of the Office of Pesticide Programs basically register and re-register pesticides and do hazard assessments, cost-benefit analyses, and risk assessments of pesticides. This type of lopsided bureaucratic arrangement is a consequence of risk-based decision making, wherein the major question being asked is "How much of a hazardous activity will be permitted?" rather than "Are there alternatives?" However, it is not difficult to imagine bureaucratic arrangements that would provide for collaboration among different specialties and for collaboration with knowledgeable members of the public.

Realistically, no alternatives assessment will consider all the alternatives that in fact are feasible and reasonable. We can only insist that the range of alternatives be broad, and that no reasonable alternatives be kept off the table simply because they make some powerful entities uncomfortable, threaten a certain corporation's profits, or require consumers to take a little more time or care.

Barrier 8: Alternatives assessment often results in a call for changes that will be resisted by powerful sectors.

Societies are capable of major social changes. For instance, many industrialized societies have moved from being largely rural to being primarily urban. Making the United States battle-ready for World War II involved enormous changes in production and job markets. We have changed from being a society that had no televisions to one that relies heavily on television. Most of the changes that societies make, however, are supported or initiated by corporations and by the wealthy and powerful.

Changes on behalf of non-human species, changes on behalf of future generations, changes on behalf of those who are ill or economically disadvantaged, and changes that involve restraint on powerful technologies or heavy consumption are resisted.

There are probably no stronger, higher, or wider barriers to alternatives assessment than these two: it considers alternatives that involve changes in business as usual, and it is a public process. Those who profit from continuing business as usual sometimes have billions of dollars to lose if changes are made. This places them in the position of wanting to prevent alternatives-assessment processes from taking place. Similarly, those government bureaucrats who lack either imagination or courage try to prevent the consideration of particular alternatives that might anger powerful interests or businesses. And finally, citizens and workers who believe in propaganda dichotomies such as in "jobs versus environment" may fear change.

To counter this barrier of fear, it is imperative that those who promote alternatives assessment in whatever form also

• try to articulate the social, environmental, and economic benefits that would be brought by possible alternatives
• describe some attractive mechanisms by which such alternatives can be considered and by which change can actually be implemented
• appeal to broad sectors of the society with proposals for alternatives assessment and change, because change will be resisted by those who are benefiting from the status quo.

Barrier 9: Cooperation by citizen groups with risk assessment legitimizes the process.

While a few citizen groups have actively pursued and used risk assessment, as noted in the Alar example above (barrier 4), many have simply cooperated with the process. These groups faithfully comment on risk assessments, critique risk assessments, submit data that should logically alter a risk assessment, suggest different numbers from scientific literature for the risk assessment, and/or hold up specific risk assessments as showing that there really is a threat (e.g., with dioxin). When questioned about why they do this, different groups will cite various reasons. Their arguments for cooperating with risk assessment might include these:

"It's the only game in town."

"We won't look credible if we don't talk this language."

"By working on the risk assessment, we can make it better."

"If we don't change this specific risk assessment, it will be adopted unchallenged."

"We asked authorities whether these chemicals were safe for us to drink (breathe, etc.) and so now they're doing this risk assessment for us."

"This risk assessment will surely show there is a hazard." (This statement implies that the only reason it hadn't been addressed earlier, is that the evidence of it being a "hazard" hadn't been convincing enough).

Once the risk-assessment process is entered, it is generally a major burden on a citizen group's time, attention, research efforts, staff expertise, and money. At some point (it took me about three years), individuals in many groups learn how infinitely malleable risk assessments are, how they are inevitably politically manipulated, how difficult they are to influence or challenge, how removed they are from the general public, and how scientifically absurd they are. But in the meantime, these groups legitimize the process by participating in it. If the risk assessment is not looking at a reasonable range of alternatives (and most aren't), the group is left discussing the potential harms of lousy behaviors and nothing else. If the risk assessment is designed to indicate an acceptable level of harm (and most are), the group is left to argue that a lower amount of damage is an "acceptable" level of harm. Meanwhile, the group is diverted from delivering the fundamental message that no amount of harm is acceptable if it is being caused unnecessarily.

In 1993, the Environmental Working Group, headquartered in Washington, D.C., chose for strategic reasons to undertake a campaign to develop "better" risk assessments for pesticide residues on foods eaten by young children.[4] The EWG was aware that pesticide risk assessments generally assume the exposed person to be an adult, and that for a variety of reasons children are more vulnerable to many pesticides (Wiles et al. 1998; Wiles and Campbell 1993). Therefore, the EWG called for re-doing food residue pesticide risk assessments to take children and infants into account. Its purpose, ultimately, was to reduce the use of pesticides in the United States, because any such revised pesticide residue limits would surely be lower than then permitted amounts. However, in choosing to pursue "better" risk assessments of pesticides, the EWG legitimized the process of determining "safe" or "insignificant" amounts of pesticides on foods that infants will eat. Therefore, it accepted a concept called "negligible risk," in which a certain amount of a pesticide in food is acknowledged to pose some potential harm but the risks are purportedly shown to be minor or "negligible." The EWG knew that the determination by agencies that a pesticide causes "negligible risk" did not and would not, in the foreseeable future, consider such effects caused by toxic "inert" ingredients in the full formulations, concurrent exposures to other toxic substances, chemical sensitivities of certain individual people, interactions of the pesticide with medical drugs that people may be taking, or such troublesome effects as endocrine disruption and immune suppression. Nevertheless, the EWG choose to call for better risk assessments.

By choosing to pursue "better" risk assessments of pesticides, the EWG legitimized the entire process of designing "acceptable" pesticide contamination levels one pesticide ingredient at a time.

The EWG legitimized a process that only the chemical companies and pesticide industry have the money, time, power, and interest to seriously influence. Monsanto, for instance, is not going to sit idly by while one of its money makers is presented realistically in a risk assessment by some agency.

The EWG legitimized a process that is always stalled when the agricultural community convincingly lodges its inevitable claim that limitations of that pesticide will mean financial ruin, crop failure, higher food prices, and/or diminished capability to compete internationally. Ultimately, it legitimized a process by which infants, farm workers, other humans, and all living species are poisoned unnecessarily.

Meanwhile, this strategy did not encourage fundamental outrage at a number of realities:

• We are feeding any pesticides to our children, even though alternatives to pesticide use exist.

• These toxic chemicals are being put on our food and are being intentionally spread around in fantastically complex ecosystems about which we understand next to nothing.

• It is absurd to pretend that we will ever know a "safe" level of pesticides for infants, children, or adults.

• Our society is justifying and defending extremely poor behaviors through risk assessment rather than challenging these poor behaviors through alternatives assessment.

The pesticide-reform movement began in fundamental outrage as expressed eloquently by Rachel Carson in *Silent Spring.* Risk assessment diverted most of this outrage from the target of pesticides and pesticide manufacturers into the process of risk assessment. Pesticide-reform groups have expended much of their efforts during the past 15 years on risk assessments. These groups have tried to make risk assessments of pesticides more accountable. They have tried to show that risks are worse than those depicted in the risk assessments. They have had to defend their opposition to pesticide spraying against risk assessments that imply the pesticides that will be used pose only "negligible risk." They have had to defend against the charge that they, the victims of pesticide exposure, are being "emotional." Meanwhile, these citizens are having to pay for preparation of many of those risk assessments with their tax dollars.

All the while, some farmers have been showing that agriculture proceeds well without pesticides, some county road managers are maintaining roadsides without any pesticides, and some school systems are spraying no pesticides in the children's play areas. The use of most pesticides hasn't even been necessary.[5]

The EWG has also called for establishing "standards" for dumping industrial hazardous wastes into fertilizers and farm land (Savitz et al. 1998). The group proposes that addition of toxic industrial waste not be allowed if it "would result in a net increase of toxics in the soil over a 40-year or longer time period." This fails to question the necessity of using farmland and fertilizers at all as hazardous-waste depositories, and it bypasses the reality that relatively little is known about the manner in which toxic chemicals, either singly or in combination, interact with healthy soil.

Too few groups consider either the larger implications of legitimizing risk assessment or their potential to demand alternatives assessments. To cooperate with risk assessment is to provide one more barrier to the adoption of alternatives assessment as a standard process within our society.

Box 17.1
Policy Statement on Pesticides by California Indian Basketweavers Association

The California Indian Basketweavers Association is opposed to the use of pesticides. We have adopted this position for the following reasons:

• The web of life that connects all living things is harmed when poisons are applied to our environment.
• The biological diversity of our forests and wetlands is diminished when pesticides are used to eliminate plants that do not have commercial value.
• Many of these same plants provide us with our foods and teas, are used in baskets and for healing, ceremonial and other traditional purposes. When we harvest and use these plants, or take fish and game, we want to know that they are free of poisons. We want the assurance that we are not endangering our health or that of our children and unborn generations.
• Pesticides contribute to the poisoning of water tables and watersheds and the destruction of fisheries.
• The licensing and regulation of pesticides favors pesticide manufacturers and users over public health and environmental well-being. The long-term effects of pesticides now in use are not known. There is mounting evidence that pesticides are contributing to an increase in human cancers and to reproductive disorders throughout the animal kingdom.
• Timber can be grown profitably without the use of pesticides. The hand labor involved in site preparation and thinning can be a source of forestry jobs at a time when these are badly needed. The Hupa Tribe in northwest California manages a profitable timber industry on tribal lands, where pesticides were banned in 1978.
• We condemn the policy of acceptable risk, which maintains that there is an acceptable level of human suffering and environmental degradation that can be balanced out by the benefits of using pesticides. The cost of pesticide use to people, wildlife and ecosystems is immense, often personal and tragic, and can never be justified by economic gain.

Adopted by CIBA Board of Directors. March 5, 1994.

source: California Indian Basketweavers Association, Nevada City

18

Forces for Alternatives Assessment

Most members of the public value information, assessment of risks, reasonableness, innovation and personal morals, feelings and vision. Alternatives assessment is based on all of these.

You have a lot going for you when you call for the establishment of an alternatives-assessment process, regardless of the setting. Specifically, you most likely live in a society that believes in and values information, consideration of risks, reasonableness, innovation, personal morals, feelings, and vision.

You can and should draw on these beliefs and values to help you. Base your strategies on them. Remind both those whose destructive behaviors you are challenging and those who join you of the power of these beliefs. Challenge anyone to go against these beliefs and values.

1. Most societies value information highly.

Most societies believe that people have a right to all relevant information. We often fail to use full information in public decision making, but you would be hard put to find a decision maker who would claim that hiding or omitting information was a virtue. That is, you could ask a decision maker "Do you think you should consider reasonable options in making this decision, or do you think you ought to exclude consideration of reasonable options from your decision making?" Risk assessment of bad options keeps the most essential information (i.e., the benefits of options) hidden. To put this another way: What's the alternative to considering alternatives? It is to make decisions without assessing alternatives. That's not a stance most people want to defend. Your job is to place the issue clearly in these terms and thereby force decision makers to provide, and consider, all relevant information.

2. Most people believe in assessing risks.

There is no need to oppose assessing risks. Alternatives assessment uses science, logic, and common sense to compare risks, as appropriate. (Sometimes a qualitative assessment of risks is enough.) However, in contrast to risk assessment, assessment of a range of alternatives can incorporate diverse values, diverse ways of observing what is happening in the world and in communities, and diverse feelings, desires, and hopes.

The most common skeptical response I get when describing alternatives assessment is "But don't we need to assess risks?" My answer is "Yes! But, assess the risks and benefits of a range of options."

3. Most people respond favorably to reasonable proposals.

Alternatives assessment is eminently reasonable to almost everyone, because it is simply thinking about alternatives before making important decisions. It makes sense to people that they get to choose their own friends from among options. It makes sense to them that, when making a decision about buying a car, they are allowed to consider not only the different kinds of cars available but also whether they even need a car. Most people would at least claim to believe in looking at the pros and cons of different options in their own lives before they make important decisions.

Most people can see that it is unreasonable to let someone decide whether they should ingest dioxin or lose spotted owls from the face of the Earth when the only options that have been considered are chlorine-based pulping technologies and the logging of the last old-growth forests in the Northwest.

Think about who doesn't want options to be looked at, and why these people don't want options to be looked at. Reveal such people for their unreasonableness and lack of responsibility.

4. People like to feel good about themselves.

Industrialized and capitalist societies take pride (deservedly or not) in encouraging innovation and invention. People like to feel they are "breaking new ground." When a paper mill switches to chlorine-free or tree-free processes, for instance, it feels good to the workers, the management (usually), the surrounding community, and the world. They get to feel clever.

In 1993, Lars Wisin, spokesperson for the Swedish flooring company Tarkett, held a press conference about Tarkett's decision to phase out its use of vinyl chloride (PVC) to make flooring. Wisin said "Many of you from the press and industry, certain clients, suppliers of raw materials, etc. will say that Tarkett is giving in [to] the requirements of the environmental movement. Yes, we are of course. But I don't see any fault in that. . . . Personally I think, as a member of society, that it is important to do what we can to save our environment. What I can do in that direction I will do."[1]

Many people personally think the way Lars Wisin thinks. Alternatives assessment increases the likelihood that a good alternative will be chosen, which increases the likelihood that people will feel good about themselves and about the way they are behaving. That's a large force to tap into.

5. Most people place a high value on morals, feelings, and positive vision.

Those pushing a hazardous activity don't have much to offer people in the way of morals when they talk about causing "acceptable" damage while they haven't even considered better alternatives or the possibility that they could be restoring health and the environment instead of damaging it further.

You, on the other hand, can take the moral high ground when you call for alternatives assessment. You are advocating that the best possible behaviors be considered. The ground you are standing on is "Care for others as you would have them care for you."

Risk assessment erases feelings about having clear water. Risk assessments don't describe the personal terror and suffering of cancer. They don't talk about the wonder of a mountain morning. They don't show photos of shorebirds born without eyes. They don't talk about people's despair at how open spaces are disappearing. Instead, they transform all these realities into numbers, so that monstrous decisions can be made "objectively," without taking feelings into account.

Alternatives assessment, on the other hand, arises out of feelings about water, mountains, children, and alternatives to despair or cancer. Therefore, feelings are your strong suit. While remaining accurate, be as explicit, poetic, eloquent, moving, personal, and passionate as you can be.

The only visions that risk assessments put forth are negative ones: scenarios of one risk after another. Alternatives assessment, on the other hand, doesn't merely look at the potential for death, extinction, destruction, or ill health. Alternatives assessment also looks at the possibility for goodness, recovery, care, change for the better, abundance of wildlife, or the return of a stream from death to life. All these positive visions are deeply embedded in emotion. Most people want to feel hopeful when given the space to do so. And when they do feel hope, people can be extraordinarily powerful forces for positive change.

Alternatives assessment offers the vision of humans behaving as decently as possible toward all of the Earth and its future. Most of us deeply want and need such a vision. Through alternatives assessment, you help people have hope.

Speak plainly.

Notes

Chapter 1

1. Particulates are small particles suspended in the atmosphere, especially pollutants.

2. An organic compound is a chemical that contains carbon and some other atoms. A chlorinated organic compound is a chemical that contains both carbon and chlorine; it may also contain other atoms, such as hydrogen and oxygen. Chlorinated organic compounds, sometimes called organochlorines, are rarely made in nature. Most of them are produced by humans, and most are toxic in some way to most living organisms.

3. The National Research Council is an organization of scientists, engineers, and other professionals serving as volunteers on approximately 900 study committees. It serves as an independent advisor to the federal government on certain scientific and technical questions deemed to be of national importance (Jaszczak 1997). Its address is 2101 Constitution Ave NW, Washington, DC 20418.

4. A gram is about a thirtieth of an ounce; a microgram is a millionth of a gram. A liter is approximately equal (depending on the liquid) to a kilogram, or 2.2 pounds; it is slightly larger than a U.S. quart. A deciliter is a tenth of a liter. Ten micrograms of lead per deciliter of blood is approximately 100 parts of lead per billion parts of blood.

5. Polychlorinated biphenyls (PCBs) are a group of human-made chemicals that consist of rings of carbon atoms to which are attached chlorine and hydrogen atoms. They were once produced industrially for use in plasticizers, adhesives, hydraulic fluids, and electrical transformers. They persist in air, soil, water, and sediments, and can now be found in the tissues of organisms throughout the Earth, even in polar regions far from industrial sources (Tanabe 1988). They are attracted to fats (i.e., are lipophilic) and become increasingly concentrated as they are transferred up through food webs. PCBs are transferred to developing embryos through the placenta in humans and to nursing infants through fat-rich human milk. PCBs cause skin diseases, reduced birth size, altered motor behavior, and reductions in short-term memory and certain types of learning (Jacobson et al. 1990).

6. Many ranchers in the western U.S. are convinced that prairie dogs compete with cattle for forage (edible plants), although scientific studies indicate that forage

competition is minimal and that plant regrowth after prairie-dog foraging is of higher nutritional quality, preventing losses in cattle weight (O'Melia et al. 1982; Hansen and Gold 1977). However, because of an undocumented 1901 Bureau of Biological Survey statement that forage competition was major (Merriam 1901), most prairie-dog towns on grasslands of the West have been poisoned and eliminated. As a result, black-footed ferrets, which eat only prairie dogs, are among the most endangered mammals on Earth (Kenworthy 1992). Some ranchers privately acknowledge that prairie dogs are compatible with cattle on their ranch, causing little or no impact, but hesitate to differ publicly with the multi-generational habit of eliminating prairie-dog towns upon which ferrets depend (personal communication, Jonathan Proctor, Prairie Dog Ecosystem Campaign Coordinator, Predator Project, Missoula, Montana, July 26, 1999).

7. Examples of structural bases of resistance include current corporate decision-making processes, separation of regulatory specialists from alternative-technology specialists in government agencies, and legal privileges that protect business as usual.

Chapter 2

1. The exposed person is almost never considered to be a sensitive embryo; it is generally considered to be an adult male.

2. Some people have become extremely sensitive to certain chemicals.

3. In a speech a year earlier, soon after he had been appointed EPA Administrator, Ruckelshaus (1983) admitted some of the uncertainties that allow people to produce wildly divergent risk assessments.

4. Narcosis is the state of altered consciousness, unconsciousness, or arrested activity produced by narcotic drugs or other chemicals. Ppm stands for parts per million.

5. "2,4-D" is a shorthand name for 2,4-dichlorophenoxyacetic acid, an herbicide that was produced by the Vertac Chemical Company. It is one of the two herbicides in Agent Orange, which the U.S. military sprayed on Vietnam.

6. "Superfund" refers to a 1980 act of Congress that instructed the EPA to find uncontrolled chemical dump sites, identify the polluters, and make them pay for cleanup.

Chapter 3

1. Polycyclic aromatic hydrocarbons are compounds with three or more closed carbon-based rings. During combustion, linear carbon-hydrogen molecules often curl into ring molecules. PAHs—the first chemical compounds shown to be carcinogenic—include benzene and benzo(a)pyrenes (Hart 1984).

2. A rem (roentgen-equivalent-man) is an estimate of the damage caused to a specific human body part such as bones or lungs by a dose of radiation. Different radioactive elements cause different amounts of damage to different tissues and

organs in living organisms. Calculation of a rem is therefore based on assumptions about the actual damage or accumulation of radioactivity in body parts. The Department of Energy notes: "These assumptions result in some uncertainty, but this approach allows more meaningful measurements than measuring energy levels from a source. Since a single radiation dose has different effects on different body organs, it is not easy to predict what effect a given dose will have on a person's health." (USDOE 1995)

3. The International Commission on Radiological Protection provides recommendations on radiation protection. It has official relationships with the World Health Organization and the International Atomic Energy Agency.

4. The Atomic Energy Commission, for instance, produced the famous Rasmussen report of 1974 to assure the public of nuclear reactors' safety. They estimated that an accident with significant core melting but no fatalities would occur once in 20,000 to 200,000 reactor years. An accident of this type occurred at Three Mile Island after fewer than 500 reactor years (Mazur 1992). Although only 135 out of 80,000 annual dose readings of workers at the Oak Ridge National Laboratory exceeded the present occupational standard for ionizing radiation, a 1991 study found their leukemia death rate is 63% above the national average for US white males (Wing et al. 1991). Their average exposures were hardly above usual background levels of ionizing radiation. This radiation-cancer mortality did not appear until after 1977, 34 years after ORNL began operation (in 1943). This study questions the existing standards for occupational exposure to low levels of ionizing radiation (Wing and Shy 1992). Scientists studying the *Exxon Valdez* oil spill of 1989 found, years later, effects on animals that no one had foreseen, such as swelling of the sheaths around the nerve cells in the brains of seals, which is the damage seen in people who die from solvent abuse. Sea otters were expected to die from hypothermia when oil clogged their fur, but large numbers of otters that were cleaned died from emphysema after breathing in toxic chemicals from the oil, and from liver and kidney failure after ingesting oil (Pain 1993).

Chapter 4

1. The stratosphere is the Earth's upper atmosphere, extending above the troposphere to an altitude of about 30 miles. The troposphere is the lower layer of the atmosphere, varying in height from 6 to 12 miles. Most clouds and weather conditions occur in the troposphere.

2. A neurotoxin is a substance that is poisonous to nerve cells.

Chapter 5

1. A riparian area is the land next to a river, stream, lake, or tidewater. It is a specialized habitat that can be critical for wildlife that live in the water, next to it, or even farther away on land.

2. In several western states (Washington, Oregon, Idaho, Montana west of the Rockies), aluminum processing accounts for 12–18% of electricity consumption.

3. During the 1940s, for instance, the U.S. Food and Drug Administration started using laboratory animals to determine what levels of exposure to toxic chemicals supposedly would not cause adverse effects. They began to establish NOELS (no adverse effect levels) for chemicals by determining the highest concentration of each toxic chemical that laboratory rodents could tolerate without becoming ill. They also introduced "safety factors" for translating laboratory animal NOELs into "safe" levels for humans. For instance, they divided the NOEL by ten in case a human was ten times more vulnerable to the toxic chemical than the laboratory animal. Another factor of ten was used in case certain human individuals were ten times more vulnerable than an "average" human. These 50-year old "safety factors," although arbitrarily chosen, are still used (Wartenburg and Chess 1993).

4. "Its composition warrants claims made for it" means, for instance, that if the product is to be marketed as a fungicide, its composition is such that EPA finds it plausible that it kills fungi. The EPA (which administers the registration of pesticides) does not consider whether the pesticide effectively controls the pest. An exception to this is the EPA's concern for the efficacy of pesticides used for public-health reasons—e.g., that of rodenticides (personal communication, pesticide registration expert Terry Shistar, February 18, 1996).

5. In general, plaintiffs in a disparate impact case under the Civil Rights Act "may ultimately prevail by proffering an equally effective alternative practice which results in less racial disproportionality" (*Georgia State Conference of Branch of NAACP v. State of Georgia*, 775 F.2d 1403, 1417 (11th Cir. 1985)).

6. Recommended reading regarding the personal nature of citizens' local caring about swamps and other hammered natural resources is an essay entitled "Water Songs" in Terry Tempest Williams's 1994 book *An Unspoken Hunger*.

Chapter 6

1. Wet scrubbers use water and other liquids to scavenge particular pollutants from the gases that are escaping out an incinerator stack. The resulting scrubber fluids, with the pollutants they have captured, must then be disposed of as liquid waste (effluents).

2. Fly ash is the small particles of ashes, dust, and soot carried out the stack of an incinerator or captured in a pollution control device. Toxic chemicals and heavy metals may be carried on the particles or may be part of the particles.

3. Hexachlorobenzene is a synthetic, chlorinated, carbon-containing compound (i.e., an organochlorine) that causes the type of environmental and health damage caused by other fat-soluble toxics such as DDT and PCBs. It is persistent in the fats of animals. It causes cancer and birth defects in laboratory animals. It is suspected human carcinogen. An EPA monitoring study found 98% of humans have HCB at measurable levels in their body. Most hexachlorobenzene is a waste product of the chlorine industry (USEPA 1985c).

4. While dealing with the serious dangers posed by hazardous waste incinerators, many communities do not realize that a great deal of hazardous waste is disposed in cement kilns, which are far less regulated (see chapter 8).

5. Several other mills produce secondarily unbleached recycled paper, in which chlorinated paper is recycled but is not bleached a second time with chlorine compounds.

6. There are two major types of pulp mills in the U.S.: kraft and sulfite mills. They differ structurally, in the wood they pulp (sulfite mills tend to use hardwoods, kraft mills use softwoods), in the chemicals they use (kraft mills boil the wood chips in caustic soda, sulfite mills boil the chips in sulfuric acid), and in the alternatives they have for producing light-colored paper without using chlorinated compounds. In the U.S. there are approximately 102 chlorine-using kraft mills and about 12 chlorine-using sulfite mills. Worldwide, about 78% of pulp is kraft pulp and about 8% is sulfite pulp (Kroesa 1990).

7. ISO standards are set by the International Organization for Standardization, a federation of over 100 national industrial standards bodies. Some of these bodies are heavily governmental; others (as in the U.S.) are industry organizations. ISO was founded in 1946, at which time it focused primarily on standardization of mechanical projects, e.g., the threads of screws (Roht-Arriaza 1995). The purpose of technical standardization was to help assure companies that they could buy products internationally and know that the products would "work" with their equipment. When a product is certified as meeting an ISO standard, the buyer presumably has some assurance that the product is of good quality.

8. Fumigant pesticides (i.e., pesticides in a vapor form) are designed to kill numerous life forms and to travel easily through air and soil so that they can spread between the soil particles in the land being fumigated. They are purposely applied on large areas of land (i.e., agricultural fields). The vaporous and poisonous nature of fumigants, which makes them "useful" as pesticides, also makes them dangerous. Fumigants inevitably escape into the atmosphere, easily move down into underground water supplies, and often poison humans and other non-pest species exposed to them. When methyl bromide escapes high into the atmosphere (e.g., the stratosphere), it destroys ozone molecules.

9. Nematodes are a class of worms that are parasitic in animals or plants or free-living in soil or water. Many nematodes are considered "beneficial" to agriculture; some are considered "pests." There are approximately 12,000 species of nematodes.

10. Primary treatment of municipal sewage removes only about 60% of suspended solids and 35% of biodegradable organic material. Further treatment is necessary to break down and disinfect the remaining organic matter.

Chapter 7

1. In 1994 the U.S. House of Representatives proposed a "Risk Assessment Improvement Act" (H.R. 4306). In 1995 the Senate proposed a "Comprehensive Regulatory Reform Act" (S. 43). Neither of these bills passed.

2. The problem of cumulative impacts is that the impacts of the one assessed substance or activity often can contribute to even more serious impacts when combined with other substances or activities.

Chapter 8

1. Chloroform causes cancer in humans, and can damage fetuses, the kidneys, liver, and nervous system (NJDOH 1993). While it is produced commercially for industrial purposes, it is also produced unintentionally, as when organic compounds in water react with chlorine, for instance, in chlorinated drinking water and swimming pools (Aggazzotti et al. 1993). 2,4-D (2,4-dichlorophenoxyacetic acid) is a chlorophenoxy lawn, forest, and agricultural herbicide that has been shown in a number of human studies to cause cancer and nerve damage in extremities (e.g., fingers, toes, hands, feet; Bane 1991). In laboratory animals, it has been shown to cause nerve and kidney damage, birth defects, and cancer (Bane 1991). Many CFCs used in automobile air conditioners and refrigerators are potent ozone depleters (Rowland 1989).

2. Since 1886, corporations have been accorded many of the rights of individual persons. For instance, corporate contributions to political candidates and campaigns are protected as free speech. Unannounced visits to factories by Occupational Safety and Health Administration officials to see if factories are following legal mandates that protect workers were ruled unconstitutional on the grounds that such visits violate the corporations' Fourth Amendment rights against illegal search and seizure. The legal purpose of a corporation, however, is to produce a surplus of money that can be returned to investors. This requires that costs be reduced, and one major way costs can be reduced is by externalizing them onto the environment, communities, and workers. Corporations use air, water, and community residents, for instance, as their free waste dumps for toxic chemicals.

3. Maquiladoras are foreign-owned assembly plants operating in Mexico under a special free-trade agreement between Mexico and the U.S. Most plants assemble electronic, textile, and high-tech computer goods. The Mexican government benefits from the investment in maquiladoras because they generate $3.5 billion in foreign exchange. Multinational and other corporations create maquiladoras because the corporations receive favorable tax and tariff incentives, are subject to loose enforcement of environmental regulations, and provide low wage rates. Workers in maquiladoras are paid little and most receive neither benefits nor job security. The maquiladoras are generally sweat shops with open vats of toxic chemicals, excessive heat, poor ventilation, and little or no safety equipment. Maquiladoras emit toxic wastes into the surrounding areas (Wright 1993).

4. Listed in CFR 40, Part 82, Appendix A to Subpart A of the Clean Air Act. Class I ozone-depleting substances are those that have an "ozone-depleting potential" or ODP of 2.0. This means the substance destroys many ozone molecules in as short period of time or a moderate amount of ozone molecules over a protracted time.

5. Citizens do not gain money when they win a citizen suit against the federal government. Instead, the benefits of such suits are non-monetary and public in nature (Axline 1991).

Chapter 9

1. Most of the information in this chapter is from *The Dairy Debate*, a 1993 book edited by William Liebhardt under the auspices of the Sustainable Agriculture Research and Education Program of the University of California at Davis. Where I used information from other sources, I identify these sources.

2. Cattle produce bovine growth hormone naturally in their pituitary gland and secrete it into their bloodstream. This hormone affects muscle, liver, bone, and mammary-gland cells. It stimulates growth and milk production; increases a cow's metabolism; and stimulates bone growth in young animals, muscle growth in adults, and food intake in all animals. In dairy cows, it increases the amount of milk that is produced per unit of food.

3. Genetic engineering is controversial because it involves removing a gene from one organism in which it had evolved and putting it in another organism, which will then function differently. We may be intending to cause the genetically altered plant or animal to function differently in one way we want. In fact, however, the organism may function differently in the environment in ways which we had not counted on and which may be harmful. Additionally, the gene may be passed to organisms we had not intended it to reach (see, e.g., Hileman 1995). For instance, an herbicide-resistance gene inserted into oilseed rape (a crop plant in the canola family) was found to have transferred to a weedy species of field mustard (also in the canola family) growing nearby (King 1996).

4. Approximately 10% of Wisconsin's dairy farmers use at least some techniques of rotational grazing to provide the primary source of feed during grazing months (Jackson-Smith and Barham 1994).

5. See Hassanein and Kloppenburg 1995 for an interesting discussion of rotational-grazing farmers in Wisconsin, many of whom have gained a sense of themselves as producers of local knowledge because they are working out among themselves the techniques that work for their farms rather than depending on university or on agribusiness companies for direction.

Chapter 10

1. Here I paraphrase and amplify some points made by the economist Holly Stallworth (1995).

2. This method of mining extracts ounces of gold from tons of soil by pouring cyanide on heaped-up soil. The cyanide then causes environmental contamination.

Chapter 11

1. A copy of the NEPA regulations, in booklet form, can be obtained by writing to the following address: Council of Environmental Quality, Executive Office of the President, 722 Jackson Place NW, Washington, DC 20503.

2. Sometimes a lawyer will agree to work on such a case *pro bono*.

3. An environmental assessment (EA) is a NEPA document that involves a less rigorous process than an EIS does. Technically, an EA is used to determine whether significant environmental impacts may occur. If the answer is affirmative, an EIS must be prepared. In reality, numerous projects that could cause severe environmental impacts slide under the wire with only an EA.

4. An article titled "Public-Interest Pretenders" (*Consumer Reports* 1994a) helps consumers to separate true "citizen" and "consumer" groups from phony ones funded by industry.

5. "Mitigation" might involve putting a liner under a hazardous waste spill, or trying to build a new wetlands in an old agricultural field after a functioning wetlands has been destroyed.

6. Chemical producers and chemical users would be those industrial segments required to comply with the Occupational Safety and Health Administration Process Management Safety Rule and with Section 112(r) of the Clean Air Act.

7. Material for the following section was provided by Joel Tickner, a doctoral candidate in the Work Environment Program of the University of Massachusetts at Lowell.

8. The Toxics Release Inventory (TRI) is a federal right-to-know mandate, key provisions of which were passed by one vote (212–211) in 1985. Chemical companies fought it, and the EPA opposed it. TRI requires certain large manufacturers to report publicly their environmental releases (to land, air, and water) and off-site transfers of 600 listed toxic chemicals. Although it is a landmark law, it has limitations: Most toxic chemicals are not on the list to be reported. Non-manufacturers and small firms are exempt. Facilities generally estimate, rather than measure, their releases. Approximately one-third of covered facilities fail to report. Chemicals in products (as opposed to wastes) are not reported (Working Group 1993). As much as 95% of toxic emissions are not covered by the law (Sheiman 1991).

9. The Montreal Protocol is binding on the 104 countries ("parties") that have signed it. Amendments that are made at Montreal Protocol meetings are binding only on those countries that ratify them in their country. Adjustments in schedules for phaseouts of ozone depleting chemicals, on the other hand, are binding on all the parties to the protocol.

10. The half-life of a toxic chemical or metal is the time required for the concentration of a substance to diminish to half of its original concentration, or presence. A half-life in an animal of 8 weeks, for instance, would mean that if an animal at one point of time were carrying 4 micrograms per kilogram body weight of a toxic chemical in its body, 8 weeks later it would still be carrying 2 mg per kg of that

chemical. The other 2 mg might have been excreted from the body, or half of the chemical might have been transformed into another chemical inside the body. Of course, if the animal were exposed to additional amounts of the chemical during those 8 weeks, it would have more than 2 mg/kg in its body at the end of the 8 weeks.

11. Bioaccumulation means that the concentration of a substance in an organism's tissues has become greater than its concentration in the surrounding environment. Many persistent toxic organochlorines, for instance, bioaccumulate in an animal if they are stored in the fats of the animal rather than passing through the body. As the animal continues to eat the organochlorine-containing foods (e.g., prey), the food is burned up for energy or excreted out of the body, but the organochlorines are retained in the animal's fats, often accumulating at higher and higher concentrations.

12. Integrated pest management (IPM), in its original, intended meaning, reduces pesticide use through avoiding conditions favorable to pests, tolerating certain levels of certain pests, and utilizing non-chemical controls (Flint and Van den Bosch 1981). Pesticide manufacturers, however, like to define IPM as a reliance on a mixture of pesticides and other controls.

13. Despite having won the exemption, the university subsequently decided to move the proposed site a quarter of a mile from the original site, in order to minimize disruption of the squirrel's habitat. The federal appeals court in a 1995 decision ordered the university to conduct an Environmental Impact Analysis on the new site (Mervis 1995).

Chapter 12

1. "Cholera" is a term used to cover several diseases marked by severe upset of the gastrointestinal system. One major cholera disease involves severe diarrhea caused by bacteria in the small intestine.

2. Nuclear fuel fabrication is the production of enriched uranium that sustains chain reactions in fission reactors. Enriched uranium has more uranium 235 in it than natural uranium and is used to make nuclear weapons. Uranium 235 has a half-life of 714 million years, meaning that half the uranium 235 that is present in nuclear wastes now will be radioactive 714 million years from now. Irradiation is the bombardment of "target" materials in a nuclear reactor by neutrons to produce new, human-made radioactive materials. Targets of uranium 238, for instance, were bombarded to make plutonium. Uranium 238 has a half-life of 4.5 billion years. Chemical separation is the extraction of uranium and plutonium from spent nuclear fuel and irradiated targets.

3. The Umatilla Tribes participated in both groups.

4. Riparian habitats are the areas adjacent to and including rivers, streams, seeps, ponds, springs, and wetlands.

5. Legitimate trade secrets, legally and narrowly defined, would be exempted.

6. "Large users" are defined as those using more than 2640 pounds of hazardous chemicals a year and having ten or more employees. Medical-waste and hazardous-waste incinerators and facilities using radioactive materials are included. The hazardous chemicals covered include all those listed as hazardous in all major federal statutes.

7. The URL is www.ci.eugene.or.us/toxics.

8. Number 97-455, Supreme Court of the State of Montana 1999 MT 248.

Chapter 13

1. The ability to sue, of course, depends of having access to an attorney who cares about the case and has the skills and ability to carry the case. In some areas, public-interest environmental attorneys are scarce. Sometimes a group does not have enough money to hire an attorney, and if no attorney will carry the case pro bono then the group may in reality not have access to the courts. The Equal Access to Justice Act provides for attorneys to recover their fees from the federal government if they win a public-interest suit against the government (Axline 1991). This allows some attorneys to take public-interest environmental cases on a pro bono basis. The importance of "equal access" provisions cannot be overemphasized.

Chapter 14

1. Absolutely zero harm is essentially impossible. A closed-loop process, for instance, may involve environmental harm in producing the pipes that are used in the closed-loop process. "No harm," however, can be considered narrowly. For instance, a farmer who does not use the pesticide methyl bromide, but instead builds up a healthy, biologically diverse soil that contains predators of the crop plant's pests and that nourishes healthy crop plants so they resist pests, is causing no pesticide harm. However, the crop may have replaced highly diverse, even critical, wildlife habitat, which is a form of harm.

Chapter 15

1. The U.S. Animal Plant and Health Inspection Service once stated this in a risk assessment for its proposed aerial spraying of carbaryl for gypsy moth control. I later found a report of a human experiment demonstrating that more than 70% of the carbaryl is absorbed through the skin (Webster and Maibach 1985).

Chapter 16

1. *Rachel's Environment & Health Weekly* is published weekly by the Environmental Research Foundation. For subscription information, write to the foundation at

P.O. Box 5036, Annapolis, MD 21403-7036, phone (410) 263-1584, fax (410) 263-8944, or email erf@rachel.org.

Chapter 17

1. See, e.g., Senate Bill 343, the "Comprehensive Regulatory Reform Act of 1995." (This bill was not passed.)

2. "Externalities" refers to the transfer of the true "costs" of production onto the community and environment through, e.g., depositing toxic waste, sedimentation of streams, spills, depleting a commonly shared resource such as water, or contributing to ill health.

3. Source of staffing-level estimates: personal communication from Dwight Welch of EPA, July 24, 1998.

4. In October 1998, the EWG resigned from the Pesticide Tolerance Reassessment Advisory Committee and the Pesticide Policy Dialogue Committee because the Clinton Administration had not taken any tangible action to protect children from pesticides (Cook 1998).

5. Interestingly, in April 1999 all seven environmental, consumer, and public interest members resigned from a federal advisory panel that had been formed to advise the EPA on implementing the 1996 Food Quality Protection Act, which was supposedly enacted to allow "only" those pesticide residues on food that are "safe" for children (see chapter 4). The EWG had been instrumental in passage of the Food Quality Protection Act. As a reason for their resignation, the groups said that the EPA had "dithered in endless, fruitless [risk assessment] debate instead of developing a plan for banning or limiting the use of agricultural chemicals that can cause cancer, neurological damage and reproductive defects" (Claiborne 1999).

Chapter 18

1. Translation provided by Greenpeace Sweden.

References

Aggazzotti, Gabriella, Guglielmina Fantuzzi, Elena Righi, Pierluigi Tartoni, Teresa Cassinadri, and Guerrino Predieri. 1993. Chloroform in alveolar air of individuals attending indoor swimming pools. *Archives of Environmental Health* 48: 250–254.

Ames, Bruce, and Lois Gold. 1990. Too many rodent carcinogens: Mitogenesis increases mutagenesis. *Science* 249: 970–971.

APHA (American Public Health Association). 1993. Recognizing and Addressing the Environmental and Occupational Health Problems Posed by Chlorinated Organic Chemicals. Resolution, San Francisco, October 2.

Ashford, Nicholas. 1993a. *The Encouragement of Technological Change for Preventing Chemical Accidents: Moving Firms from Secondary Prevention and Mitigation to Primary Prevention*. Center for Technology, Policy, and Industrial Development, Massachusetts Institute of Technology.

Ashford, Nicholas. 1993b. Testimony on the proposed rule on risk management programs for chemical accidental release prevention (40 *CFR* Part 68). Draft, November 30.

Ashford, Nicholas, and Charles Caldart. 1991. *Technology, Law, and the Working Environment*. Van Nostrand Reinhold.

Ashford, Nicholas, and Claudia Miller. 1990a. *Chemical Exposures: Low Levels and High Stakes*. Van Nostrand Reinhold.

Ashford, Nicholas, and Claudia Miller. 1990b. *Chemical Sensitivity*. Prepared for New Jersey Department of Health; distributed by National Center for Environmental Health Strategies.

Axline, Michael. 1991. *Environmental Citizen Suits*. Butterworth Legal Publishers.

Bailey, Jeff. 1992. Dueling studies: How two industries created a fresh spin on the dioxin debate. *Wall Street Journal*, February 20.

Bane, Gwen. 1991. 2,4-D. *Journal of Pesticide Reform* 11 (3): 21–25.

Banks, Jonathan. 1995. Re: MBTOC. Memorandum to Methyl Bromide Technical Options Committee, United Nations Environment Programme, August 24.

Belsky, A. J, and D. M. Blumenthal. 1995. *Effects of Livestock Grazing on Upland Forests, Stand Dynamics, and Soils of the Interior West*. Oregon Natural Resources Council.

Billings, W. 1990. *Bromus tectorum*, a biotic cause of ecosystem impoverishment in the Great Basin. In *The Earth in Transition*, ed. G. Woodwell. Cambridge University Press.

Bionetics Research Laboratories, Inc. 1969. Evaluation of the carcinogenic, teratogenic, and mutagenic activity of selected pesticides and industrial chemicals. In *Evaluation of the Teratogenic Activity of Selected Pesticides and Industrial Chemicals* (National Cancer Institute Contracts PH43-64-57 and PH43-67-735).

Blake, Tupper, and Peter Steinhart. 1987. *Tracks in the Sky: Wildlife and Wetlands of the Pacific Flyway*. Chronicle Books.

Blaustein, Andrew, Peter Hoffman, Grant Hokit, Joseph Kiesecker, Susan Walls, and John Hays. 1994. UV repair and resistance to solar UV-B in amphibian eggs: A link to population declines? *Proceedings of the National Academy of Sciences* 91 (5): 1791–1795.

BNA (Bureau of National Affairs). 1992. More Than 2,000 Bays, Beaches Closed in 1991 Due to Pollution, NRDC Says. July 24.

Bonsor, Norman, Neil McCubbin, and John Sprague. 1988. *Kraft Mill Effluents in Ontario*. Prepared for Environment Canada.

Bowman, R., S. Schantz, N. Weerasinghe, M. Gross, and D. Barsotti. 1989a. Chronic dietary intake of 2,3,7,8-tetrachlorodibenzo-p-diosin (TCDD) at 5 or 35 parts per trillion in the monkey: TCDD kinetics and dose-effect estimate of reproductive toxicity. *Chemosphere* 18: 99–111.

Bowman, R., S. Schantz, M. Gross, and S. Ferguson. 1989b. Learning in monkeys exposed perinatally to 2,3,7,8-TCDD transmitted maternally during gestation and for four months of nursing. *Chemosphere* 18: 235–242.

BPA (Bonneville Power Administration). 1995. *Environmental Assessment: Willow Creek Wildlife Mitigation Project*. U.S. Department of Energy.

Breyer, Stephen. 1993. *Breaking the Vicious Circle: Toward Effective Risk Regulation*. Harvard University Press.

Brooke, James. 1996. Next Door to Danger, a Booming City. *New York Times*, October 6.

Brooke, James. 1997. Chemical neutralization is gaining in war on poison gas. *New York Times* , February 7.

Bruck, Glenn. 1986. *Pesticide and Nitrate Contamination of Groundwater near Ontario, Oregon*. U.S. Environmental Protection Agency.

Bryant, Bunyan, and Paul Mohai. 1992. *Race and Incidence of Environmental Hazards: A Time for Discourse*. Westview.

Cabral, J. , P. Shubik, T. Mollner, and F. Raitano. 1977. Carcinogenic activity of hexachlorobenzene in hamsters. *Nature* 269: 510–511.

Campbell, Ramsey. 1994. Researchers find "silent spring" of death in Lake Apopka. *Orlando Sentinel*, April 10.

Carlsen, Elisabeth, Aleksander Giwercman, Niels Keiding, and Niels Skakkebaek. 1992. Evidence for decreasing quality of semen during past 50 years. *British Medical Journal* 305: 609–613.

Castleman, Barry, and Grace Ziem. 1988. Corporate influence on threshold limit values. *American Journal of Industrial Medicine* 13 (5): 531–559.

Castleman, Barry, and Grace Ziem. 1989. Toxic pollutants, science, and corporate influence. *Archives of Environmental Health* 44 (2): 68, 127.

CCHW (Citizens' Clearinghouse for Hazardous Wastes). 1992. Who we are. *Everyone's Backyard* 10 (6): 2.

CCRP (California Comparative Risk Project). 1994. *Toward the 21st Century: Planning for the Protection of California's Environment.*

Christensen, Alan, Jack Lyon, and James Unsworth. 1993. *Elk Management in the Northern Region: Considerations in Forest Plan Updates or Revisions.* General Technical Report INT-303, Intermountain Research Station, U.S. Forest Service, Ogden, Utah.

CIBA (California Indian Basketweavers Association). 1994. *Policy Statement on Pesticides.*

Claiborne, William. 1999. "7 Groups Quite Food Panel: EPA Termed Soft on Pesticide Risks." *Washington Post*, April 28.

Clean Air Act. 1993. Section 611. See 58 *Federal Register* 51: 15038 (March 18).

Clinton, Bill. 1996. Remarks by the president in a radio address to the nation. Press release, Office of Press Secretary, White House, August 3.

Colborn, Theo, Dianne Dumanoski, and John Peterson Myers. 1996. *Our Stolen Future*. Penguin.

Colborn, Theo, Frederick vom Saal, and Ana Soto. 1993. Developmental effects of endocrine-disrupting chemicals in wildlife and humans. *Environmental Health Perspectives* 101 (5): 378–384.

Columbia River United. Undated. *Hanford and the River.*

Commission of the European Communities. 1991. *Benchmark Exercise on Major Hazard Analysis.*

Committee on Science, Engineering, and Public Policy, National Academy of Sciences, National Academy of Engineering, and Institute of Medicine. 1991. *Policy Implications of Greenhouse Warming.* National Academy Press.

Consumer Reports. 1994a. Public-interest pretenders. 59 (5): 316–319.

Consumer Reports. 1994b. Electromagnetic fields. 59 (5): 354.

Costner, Pat. 1990a. *We All Live Downstream: A Guide to Waste Treatment That Stops Water Pollution.* Waterworks.

Costner, Pat. 1990b. *We All Live Downstream: For Everyone Who Wants Clean Water.* Waterworks.

Costner, Pat. January 29, 1992. The incineration of dioxin in Jacksonville, Arkansas: A review of the trial burns at Vertac Site Contractors and related air monitoring at Vertac Site Contractor's mobile incinerator. Unpublished paper, Greenpeace USA.

Costner, Pat. 1994. The incineration of dioxin-contaminated wastes in Jacksonville, Arkansas: A review of 1992–1993 air monitoring data. Unpublished paper, Greenpeace USA.

Costner, Pat, and Joe Thornton. 1989. *We All Live Downstream: The Mississippi River and the National Toxics Crisis*. Greenpeace USA.

Costner, Pat, and Joe Thornton. 1990. *Playing with Fire: Hazardous Waste Incineration*. Greenpeace USA.

Council on Environmental Quality. 1992a. 40 *CFR* 1502.14.

Council on Environmental Quality. 1992b. *Regulations for Implementing the Procedural Provisions of the National Environmental Policy Act*. 40 *CFR* 1500–1508.

Cranor, Carl. 1992. The accuracy and social benefits of expedited risk. Presented at Policies, Programs, and Public Participation: Environmental and Occupational Health in the Emerging Market Economies and Democracies of Central and Eastern Europe, third annual symposium, Pultusk, Poland.

Cremlyn, R. 1991. *Agrochemicals: Preparation and Mode of Action*. Wiley.

CRITFC (Columbia River Inter-Tribal Fish Commission). Undated (1993). *A Fish Consumption Survey of the Umatilla, Nez Perce, Yakima and Warm Spring Tribes of the Columbia River Basin*.

CTUIR (Confederated Tribes of the Umatilla Indian Reservation). 1995. *Scoping Report: Nuclear Risks in Tribal Communities*.

CWWG (Chemical Weapons Working Group). 1994. The citizens' solution for chemical weapons disposal. Unpublished paper.

Daily, Gretchen, and Paul Ehrlich. 1992. Population, sustainability, and Earth's carrying capacity. *BioScience* 42 (10): 761–771.

Daly, Helen. 1990. Reward reductions found more aversive by rats fed environmentally contaminated salmon. *Neurotoxicology and Teratology* 13: 449–453.

Davies, Kert. 1996. Florida's Unsafe Tomatoes. Draft for Environmental Working Group, Washington, D.C.

Davis, Devra Lee. 1993. Medical hypothesis: Xeno-estrogens as preventable causes of breast cancer. *Environmental Health Perspectives* 101: 372–377.

Dewailly, Eric, John Jake Ryan, Claire Laliberte, Suzanne Bruenau, Jean-Philippe Weber, Suzanne Gingras, and Gaetan Carrier. 1994. Exposure of remote maritime populations to coplanar PCBs. *Environmental Health Perspectives* 102, suppl. 1: 205–209.

Dockery, Douglas, C. Arden Pope III, Xiping Xu, John Spengler, James Whare, Martha Fay, Benjamin Ferris, Jr., and Frank Speizer. 1993. An association between air pollution and mortality in six U.S. cities. *New England Journal of Medicine* 329: 1753–1759.

Douglass, Frederick. 1857. The significance of emancipation in the West Indies. In*The Frederick Douglass Papers. Series One. Speeches, Debates and Interviews. Vol. 3*, ed. J. Blassingame. Yale University Press, 1979.

DSR PAC (Defense Subsistence Region Pacific). 1994. *Pacific Surface Initiative and Produce Shipment Losses to Guam.*.

Dunn, Julie Anton. 1995. *Organic Food and Fiber: Analysis of 1994 Certified Production in the United States*. Agriculture Marketing Services, Transportation and Marketing Division, U.S. Department of Agriculture.

Ecology and Environment, Inc. 1997. *Pre-Trial Burn Risk Assessment for the Proposed Umatilla Chemical Demilitarization Facility Hermiston, Oregon.* Contract 64-93.

Endometriosis Association. 1993. Endometriosis, disease affecting five and a half million U.S. and Canadian women, linked to dioxin in new studies. Press release, Endometriosis Association, November 5..

ENDS (Environmental Data Services Ltd.). 1989. Report 177, October.

Engelman, Robert, and Pamela LeRoy. 1993. *Sustaining Water, Population and the Future of Renewable Water Supplies*. Population Action International. Cited in Rapaport 1995.

Fein, Greta, Joseph Jacobson, Sandra Jacobson, Pamela Schwartz, and Jeffrey Dowler. 1984. Prenatal exposure to polychlorinated biphenyls: Effects on birth size and gestational age. *Journal of Pediatrics* 105: 315–320.

Ferraro, Richard (Appeal Deciding Officer, Pacific Northwest Region, U.S. Forest Service), letter to Ochoco Resource and Recreation Association, July 25, 1994.

FIFRA (Federal Insecticide, Fungicide and Rodenticide Act). 7 U.S.C.A. § 136 et seq.

Fischer, Carolyn, and Carol Schwartz, eds. 1996. *Encyclopedia of Associations*. Gale Research.

Flint, Mary Louise, and Robert Van den Bosch. 1981. *Introduction to Integrated Pest Management*. Plenum.

GATT (General Agreement on Tariffs and Trade). 1993. General Agreement on Tariffs and Trade Sanitary and Phytosanitary Standards Agreement of 1993, paragraph 20.

GATT. 1994. *GATT Activities*. Geneva.

Gergen, Peter, Daniel Mullally, and Richard Evans. 1988. National survey of prevalence of asthma among children in the United States, 1976 to 1980. *Pediatrics* 81: 1–7.

Gillmore, Heath. 1993. Effluent threat to city beaches. *Sydney Morning Herald*, August 29. Cited in Rapaport 1995.

Gofman, John. 1990. *Radiation-Induced Cancer from Low-dose Exposure: An Independent Analysis*. Committee for Nuclear Responsibility.

Gofman, John. 1995. *Preventing Breast Cancer: The Story of a Major, Proven, Preventable Cause of This Disease*. Committee for Nuclear Responsibility.

Gofman, John, and Egan O'Connor. 1985. *X-Rays: Health Effects of Common Exams*. Sierra Club Books.

Gofman, John, and Egan O'Connor. 1993. *The Law of Concentrated Benefit over Diffuse Energy*. Committee for Nuclear Responsibility.

Gómez, A. 1993. *El Injerto Herbaceo Como Método Alternativo De Control De Enfermedades Telúricas Y Sus Aplicaciones Agronomicas*. Doctoral thesis, Universidad Politécnica de Valencia.

Graham, R., C. Hunsaker, and R. O'Neill. 1991. Ecological risk assessment at the regional scale. *Ecological Applications* 1 (2): 196–206.

Greenpeace. 1994a. *The Greenfreeze Story: A Story of a Solution Which Governments Ignored and Industry Tried to Stop*. Canonbury Villas, London.

Greenpeace. 1994b. *Money to Burn: The World Bank, Chemical Companies and Ozone Deletion*.

Guillette, L., T. Gross, G. Masson, J. Matter, H. Percival, and A. Woodward. 1994. Developmental abnormalities of the gonad and abnormal sex hormone concentrations in juvenile alligators from contaminated and control lakes in Florida. *Environmental Health Perspectives* 102: 680–688.

Habicht, Henry. 1994. EPA's vision for setting national environmental priorities. In *Worst Things First?*, ed. A. Finkel and D. Golding. Resources for the Future.

Hansen, Michael. 1990. *Biotechnology and Milk: Benefit or Threat?* Consumer Policy Institute/Consumers Union.

Hansen, R., and I. Gold. 1977. Black-tailed prairie dogs, desert cottontails, and cattle trophic relations on short grass range. *Journal of Range Management* 30: 210–214. Cited in Miller et al. 1990.

Hart, Roger. 1984. The questionable practice of slash burning. *NCAP News* 4 (5): 17–21.

Hassanein, Neva. 1989. Enforcement of the Federal Pesticide Law: An Assessment of Oregon's Program. Master's thesis, Department of Public Planning and Policy Management, University of Oregon, Eugene.

Hassanein, Neva, and Jack Kloppenburg. 1995. Where the grass grows again: Knowledge exchange in the sustainable agriculture movement. *Rural Sociology* 60 (4): 721–740.

Hazardous Waste Facility Approval Board of the State of Ohio. 1984. Written opinion and final order approving application for hazardous waste facility installation and operation permit (16 February). Case 82-NF-0589.

Hileman, Bette. 1993. Concerns broaden over chlorine and chlorinated hydrocarbons. *Chemical and Engineering News*, April 19: 11–20.

Hileman, Bette. 1995. Views differ sharply over benefits, risks of agricultural biotechnology. *Chemical and Engineering News*, August: 8–17.

Holm-Hansen, O. 1990. UV radiation in Arctic waters: Effect on rates of primary production. In Proceedings of Workshop on Response of Marine Phytoplankton on Natural Variation in UV-B Flux, Scripps Institution of Oceanography. Cited in UNEP 1991.

Houck, Oliver. 1989. Hard choices: The analysis of alternatives under Section 404 of the Clean Water Act and similar environmental laws. *University of Colorado Law Review* 60: 773–840.

Hoversten, Paul. 1996. Thiokol wavers, then decides to launch. *USA Today*, January 22.

Hoversten, Paul, Patricia Edmonds, and Haya El Nasser. 1996. Debate raged night before doomed launch. *USA Today*, January 22.

Hurst, Peter. 1992. *Pesticide Reduction Programmes in Denmark, the Netherlands, and Sweden*. World Wildlife Fund International.

IJC (International Joint Commission on Great Lakes Water Quality). 1991. *Report to the International Joint Commission.*

IJC. 1992. *Sixth Biennial Report on Great Lakes Water Quality.*

IJC. 1994. *Seventh Biennial Report Under the Great Lakes Water Quality Agreement of 1978 to the Governments of the United States and Canada and the State and Provincial Governments of the Great Lakes Basin.*

Indigenous Caucus. 1994 Open Letter to National Institute of Environmental Health Sciences; NIH, Office of Minority Health Research; U.S. Environmental Protection Agency; National Institute for Occupational Safety and Health; CDC]; Agency for Toxic Substances and Disease Registry; U.S. Department of Energy; National Center for Environmental Health, CDC. At Symposium on Health Research and Needs to Ensure Environmental Justice, Arlington, Virginia.

Isaac, Katherine. 1992. *Civics for Democracy: A Journey for Teachers and Students.* Essential Books.

Jackson-Smith, Douglas, and Bradford Barham. 1994. Farm enterprise technology. In *Status of Wisconsin Farming (Special Edition): 1993 ATFFI Family Farm Survey.* Department of Agricultural Economics, Cooperative Extension, and Agricultural Technology and Family Farm Institute, University of Wisconsin, Madison.

Jacobson, Joseph, Sandra Jacobson, and Harold Humphrey. 1990. Effects of *in utero* exposure to polychlorinated biphenyls and related contaminants on cognitive functioning in young children. *Journal of Pediatrics* 116: 38–45.

Jaszczak, Sandra, ed. 1997. *Encyclopedia of Associations.* Gale.

Johansen, J. 1993. Cryptogamic crusts of semiarid and arid lands of North America. *Journal of Phycology* 29: 140–147.

Johnston, Roxanna. 1997. *Introduction to Microbiotic Crusts.* Grazing Lands Technology Institute, Natural Resources Conservation Service, U.S. Department of Agriculture.

Keiter, Robert. 1997. Greater Yellowstone's bison: Unraveling of an early American wildlife conservation achievement. *Journal of Wildlife Management* 61 (1): 1–11.

Kelly, Mary. 1994. Risk assessment goes international: A new role for trade agreements. *Global Pesticide Campaigner* 4 (1): 1, 9–11.

Kenworthy, Tom. 1992. The lesson of the black-footed ferret: Grazing and conservation compatible, preaches Wyoming cattleman. *Washington Post*, August 2.

Kerasote, Ted. 1997. *Heart of Home*. Villard.

Keystone Center. Undated. Ground rules for [1997–1998] dialogue on assembled chemical weapon assessment. Keystone, Colorado.

King, James. 1996. Agricultural ecology: Could transgenic supercrops one day breed superweeds? *Science* 274 (October 11): 180–181.

Kipp, Stephen. 1991. Evans will ask EPA for tougher dioxin standards. *Birmingham* (Alabama) *Post Herald*, October 28.

Kirschenmann, Frederick. 1998. Organic agriculture endangered. *Rachel's Environment & Health Weekly* no. 583: 1–2.

Knipling, Edward. 1997. USDA perspective on methyl bromide. Presentation at Third Annual International Research Conference on Methyl Bromide Alternatives and Emissions Reduction, Orlando, FL, November 4–6. Printed in: *Methyl Bromide Alternatives* (January): 14. Washington D.C.: U.S. Department of Agriculture.

Kroesa, Renate. 1990. *The Greenpeace Guide to Paper*. Greenpeace International.

Kruszewska, I. 1995. *What is clean production? A Greenpeace briefing on strategies to promote Clean Production*. Amsterdam: Greenpeace, International.

LaDou, Joseph. 1984. The not-so-clean business of making chips. *Technology Review* 87: 22–36.

LaDou, Joseph. 1994. Health issues in the global semiconductor industry. *Annals Academy of Medicine Singapore* 23: 765–769.

Last, John, ed. 1980. *Maxcy-Rosenau Public Health and Preventive Medicine*. Appleton-Century-Crofts.

Lee, G. Fred, and Anne Jones. 1992. *Municipal Solid Waste Management in Lined, "Dry Tomb" Landfills: A Technologically Flawed Approach for Protection of Groundwater Quality*. G. Fred Lee & Associates.

Leopold, Aldo. 1941. *Sand County Almanac*. Oxford University Press.

Leopold, Aldo. 1953. *Round River*. Oxford University Press.

Lewis, Barbara. 1991. *A Kid's Guide to Social Action: How to Solve the Social Problems You Choose and Turn Creative Thinking into Positive Action*. Minneapolis, Minn.: Free Spirit Publishing Co.

Lewis, Peter. 13 December 1994. IBM halts sales of its computers with flawed chip. *New York Times*, A1.

Liebhardt, William C., ed. 1993. *The Dairy Debate: Consequences of Bovine Growth Hormone and Rotational Grazing Technologies*. Sustainable Agriculture Research and Education Program, University of California, Davis.

London Dumping Convention. 1972. Convention on the Prevention of Marine Pollution by Dumping of Wastes and Other Matter.

Longstreth, J., F de Gruijl, Y. Takizawa, and J. van der Leun. 1991. Human health. In *Environmental Effects of Ozone Depletion: 1991 Update*, United Nations Environment Programme.

Ludwig, Donald, Ray Hilborn, and Carl Walters. 1993. Uncertainty, resource exploitation, and conservation: Lessons from history. *Science* 260 (April 2): 17, 36.

Luster, Michael, and Gary Rosenthal. 1993. Chemical agents and the immune response. *Environmental Health Perspectives* 100: 219–236.

Madany, M., and N. West. 1983. Livestock grazing—fire regime interactions within montane forests of Zion National Park, Utah. *Ecology* 64 (4): 661–667.

Maine Board of Pesticides Control. 1997. BPC passes on B.t. forage corn registrations: Growers, registrants fail to show need for technology. Quarterly electronic publication, December 30.

Makhijani, Arjun. 1992. *Mending the Ozone Hole*. Institute for Energy and Environmental Research. (MIT Press edition: 1995)

Marsh v. ONRC (Marsh vs. Oregon Natural Resources Council). 1989. 490 U.S. 360, 378.

Mazur, Allan. 1992. The hazards of risk assessment. *Chemical and Engineering News* 70 (41): 76–77, 106.

MBTOC (Methyl Bromide Technical Options Committee, Montreal Protocol on Substances That Deplete the Ozone Layer). 1994. *UNEP 1994 Report of the Methyl Bromide Technical Options Committee*.

MBTOC. 1995. *1994 Report of the Methyl Bromide Technical Options Committee*.

McCormack, Craig, and David Cleverly, U.S. Environmental Protection Agency. April 23, 1990. *Analysis of the potential populations at risk from the consumption of freshwater fish caught near paper mills*. Draft paper. Received via Freedom of Information Act request by Greenpeace.

MDEP (Massachusetts Department of Environmental Protection). 1998. *1996 TURA Information Release, Executive Summary*.

Meffe, G. 1992. Techno-arrogance and halfway technologies: Salmon hatcheries on the Pacific Coast of North America. *Conservation Biology* 6: 350–354.

Merrell, Paul. 1983. An answer to Mr. Ruckelshaus. *NCAP News* 3 (4): 12–13.

Merrell, Paul, and Carol Van Strum. 1990. Negligible risk: Premeditated murder? *Journal of Pesticide Reform* 10 (1): 20–22.

Merriam, C. 1901. The prairie dog of the Great Plains. *Yearbook of the Department of Agriculture*, pp. 257–270. Cited in Miller et al. 1994.

Merrill, Michael. 1991. No pollution prevention without income protection: A challenge to environmentalists. *New Solutions* 1 (3): 9–11.

Mervis, Jeffrey. 1995. Red squirrels 2, astronomers 0. *Science* 268: 630.

Miller, B., C. Wimmer, D. Biggins, and R. Reading. 1990. A proposal to conserve black-footed ferrets and the prairie dog ecosystem. *Environmental Management* 14: 763–769.

Miller, B., L. Gloeckler, M. Ries, B. Hankey, C. Kosary, and B. Edwards. 1992. *Cancer Statistics Review 1973–1989.* Publication 92-2789, National Institutes of Health.

Miller, B., G. Ceballas, and R. Reading. 1994. The prairie dog and biotic diversity. *Conservation Biology* 8: 677–681.

Miller-Freeman, Inc. 1994. Pulp and Paper Mills in the U.S. and Canada. Map.

Montague, Peter. 1988. Doctors, grass roots activists to meet, discuss toxic exposures. *Rachel's Environment & Health Weekly* no. 94: 1–1.

Montague, Peter. 1992a. The breakdown of morality. *Rachel's Environment & Health Weekly* no. 287: 1–2.

Montague, Peter. 1992b. New evidence that all landfills leak. *Rachel's Environment & Health Weekly* no. 316: 1–2.

Montague, Peter. 1994a. Chemicals and Health—Part 3. *Rachel's Environment & Health Weekly* no. 371: 1–2.

Montague, Peter. 1994b. The Corporation—Part 1. *Rachel's Environment & Health Weekly* no. 388: 1–2.

Montague, Peter. 1994c. Dioxin reassessed—Part 1. *Rachel's Environment & Health Weekly* no. 390: 1–2.

Montague, Peter. 1994d. Dioxin reassessed—Part 2. *Rachel's Environment & Health Weekly* no. 391: 1–2.

Montague, Peter. 1994e. Invisible killers: Fine particles. *Rachel's Environment & Health Weekly* no. 373: 1–2.

Montague, Peter. 1994f. The many uses of risk assessment. *Rachel's Environment & Health Weekly* no. 420: 1–2.

Montague, Peter. 1995. History of quantitative risk assessment. Unpublished paper.

Moran, Carolyn. 1994. Tree-free paper: Dispelling the myth. *Talking Leaves* 6 (2) 36–37.

Muir, Derek, Ross Norstrom, and Mary Simon. 1988. Organochlorine contaminants in Arctic marine food chains: Accumulation of specific polychlorinated biphenyls and chlordane-related compounds. *Environtal Science and Technology* 22: 1071–1079.

Muir, T., T. Eder, P. Muldoon, and S. Lerner. 1993. Case study. Application of a virtual elimination strategy to an industrial feedstock chemical—chlorine. Appendix B to volume 2 of *Strategy for Virtual Elimination of Persistent Toxic Substances* (Virtual Elimination Task Force to the International Joint Commission).

Müller, R., P. Crutzen, J.-U. Groob, C. Brühl, J. Russell III, H. Gernandt, D. McKenna, and A. Tuck. 1997. Severe chemical ozone loss in the Arctic during the winter of 1995–96. *Nature* 389: 709–712.

Murray, F., F. Smith, K. Nitschke, C. Humiston, R. Kociba, and B. Schwetz. 1979. Three-generation reproduction study of rats given 2,3,7,8-tetrachlorodibenzo-*p*-dioxin (TCDD) in the diet. *Toxicology and Applied Pharmacology* 50: 241–252.

Nash, Nathaniel. 1991. Unease grows under the ozone hole. *New York Times*, July 23.

National Research Council. 1993. *Measuring Lead Exposure in Infants, Children, and Other Sensitive Populations*. National Academy Press.

NCWARN (North Carolina Waste Awareness and Reduction Network.). Undated. Summary description. Unpublished paper.

New York City Department of Environmental Protection. 1994. Notice of promulgation of amendments to New York City's Community Right-to-Know Regulations.

NJDOH (New Jersey Department of Health). 1993. Hazardous Substance Fact Sheet: Chloroform.

Noss, Reed, Edward LaRoe, and Michael Scott. 1995. *Endangered Ecosystems of the United States: A Preliminary Assessment of Loss and Degradation*. National Biological Service, U.S. Department of the Interior.

NRC (National Research Council). 1984. *Disposal of Chemical Munitions and Agents*. National Academy Press.

NRC (National Research Council). 1990. *Health Effects of Exposure to Low Levels of Ionizing Radiation: BEIR V*. National Academy Press.

NRC. 1993. *Pesticides in the Diets of Infants and Children*. National Academy Press.

O'Brien, Mary. 1988. Quantitative risk analysis: Over-used, under-examined. *Journal of Pesticide Reform* 8 (1): 7–12.

O'Brien, Mary. 1990a. A crucial matter of cumulative impacts: Toxicity equivalency factors. *Journal of Pesticide Reform* 10 (2): 23–27.

O'Brien, Mary. 1990b. NEPA as it was meant to be: *NCAP* v. *Block*, herbicides and Region 6 Forest Service. *Environmental Law* 20: 735–745.

O'Brien, Mary. 1992. Selling risk assessment abroad: A public health threat to be challenged. Presentation at the 120th Annual Meeting of American Public Health Association, Washington, D.C.

O'Brien, Mary 1995. Ecological alternatives assessment rather than ecological risk assessment: Considering options, benefits, and dangers. *Human and Ecological Risk Assessment* 1 (4): 357–366.

Olexsey, Robert (Chief, TTS, TDB, ATD, HWERL, U.S. Environmental Protection Agency). 1985. Memorandum to James Berlow, Program Manager, Treatment Reduction and Recycling Program Re: Review of proposed rule, land disposal restriction framework and decision on solvent and dioxin containing wastes, October 24.

O'Melia, M., F. Knops, and J. Lewis. 1982. Some consequences of competition between prairie dogs and beef cattle. *Journal of Range Management* 35: 580.

Organization of African Unity. 1991. Bamako Convention on the Ban of the Import Into Africa and the Control of Transboundary Movement and Management of Hazardous Wastes Within Africa. Adopted 29 January 1991. Bamako, Mali.

OTA (Office of Technology Assessment, U.S. Congress). 1986. *Serious Reduction of Hazardous Waste for Pollution Prevention and Industrial Efficiency*.

Ozone Secretariat, United Nations Environment Programme. 1993. *Handbook for the Montreal Protocol on Substances that Deplete the Ozone Layer*.

Ozonoff, David. 1993. Taking the handle off the chlorine pump. Presentation at Public Health Forum organized by Boston University School of Public Health, October 5.

Pain, Stephanie. 1993. Species after species suffers from Alaska's spill. *New Scientist* 137 (1860): 5.

PANNA (Pesticide Action Network, North America). 1998. USDA Bows to Pressure on Organic Standards. Pesticide Action Network North America Update Service (http: //www.panna.org/panna/), May 18.

PNL (Pacific Northwest Laboratory) and CRCIA (Columbia River Comprehensive Impact Assessment) Management Team Representatives. 1997. Screening Assessment and Requirements for a Comprehensive Assessment. DOE/RL-96-16 (Draft). U.S. Department of Energy.

Paulsen, Monte. 1994. The cancer business. *Mother Jones*, May-June: 41.

Pease, W., J. Liebman, D. Landy, and D. Albright. 1996. *Pesticide Use in California: Strategies for Reducing Environmental Health Impacts*. California Policy Seminar.

Pesticide Chemical News. 1976. HCB found by EPA in 75 percent of mother's milk samples; pesticide use being examined. *Pesticide Chemical News* , June 30: 25–26.

Pettit, Greg. 1987. *Assessment of Oregon's groundwater for agricultural chemicals*. Oregon Department of Environmental Quality.

Picardi, Alfred, Paul Johnston, and Ruth Stringer. 1991. *Alternative Technologies for the Detoxification of Chemical Weapons: An Information Document*. Greenpeace International.

Platte, Anne. 1995. Dying seas. *Worldwatch* 8 (1): 10–19.

Predator Project. 1993. USFWS proposes to designate two reintroduced populations of black-footed ferrets as experimental, nonessential. *Predator Project Report/Alert*, May 28.

Raloff, Janet. 1997. Hanford tanks, leaks reach groundwater. *Science News* 152: 410.

Rapaport, Dave. 1995. *Sewage Pollution in Pacific Island Countries and How to Prevent It*. Center for Clean Development.

Reganold, John, Lloyd Elliott, and Yvonne Unger. 1987. Long-term effects of organic and conventional farming on soil erosion. *Nature* 330: 370–372.

Rice, Susan Dianne. 1991. Waste incinerator's location amounts to "environmental racism," suit says. *San Francisco Daily Journal*, February 7.

Rier, Sherry, Dan Martin, Robert Bowman, Paul Dmowski, and Jeanne Becker. 1993. Endometriosis in rhesus monkeys (*Macaca mulatta*) following chronic exposure to 2,3,7,8-tetrachlorodibenzo-*p*-dioxin. *Fundamental and Applied Toxicology* 21: 433–441.

Roach, S., and S. Rappaport. 1990. But they are not thresholds: A critical analysis of the documentation of threshold limit values. *American Journal of Industrial Medicine* 17 (6): 727–753.

Roberts, Leslie. 1990. Counting on science at EPA. *Science* 249: 616–618.

Roberts, Leslie. 1991. More pieces in the dioxin puzzle. *Research News* (254): 177.

Robertson, Lance. 1995. Plant emissions not "significant." *Register-Guard* (Eugene, Oregon), August 8.

Rodgers, K., and D. Ellefson. 1992. Mechanism of the modulation of murine peritoneal cell function and mast cell degranulation by low doses of malathion. *Agents and Actions* 35: 57–63.

Roht-Arriaza, Naomi. 1995. Shifting the point of regulation: The International Organization for Standardization and global lawmaking on trade and the environment. *Ecology Law Quarterly* 22: 479–539.

Rosenberg, Beth. 1995. The Best Laid Bans: The Impact of Pesticide Bans on Workers. Doctoral dissertation, University of Massachusetts, Lowell.

Rowland, F. 1989. Chlorofluorocarbons and the depletion of stratospheric ozone. *American Scientist* 77: 36–45.

Ruckelshaus, William. 1983. Science, risk and public policy. *Science* 221: 1026–1028.

Ruckelshaus, William. 1984. Risk in a free society. *Risk Analysis* 4: 157–162.

Savitz, Jacqueline, Todd Hettenbach, and Richard Wiles. 1998. *Factory Farming: Toxic Waste and Fertilizer in the United States, 1990–1995.* Environmental Working Group.

Sewell, Bradford, and Robin Whyatt. 1989. *Intolerable Risk: Pesticides in Our Children's Foods.* Natural Resources Defense Council.

Sheiman, Deborah. 1991. *The Right to Know More.* Natural Resources Defense Council.

Sher, Vic. 1987. Pests, poisons, and power: The constitutional implications of state pest eradication projects in California. *Journal of Environmental Law and Litigation* 1: 89–106.

Shirley, Christopher. 1993. Milking for money or for profit? *New Farm* 15 (6): 31–34.

Sierra Club Legal Defense Fund. 1993. Comments on DSEIS. In re: Draft Supplemental Impact Statement on Management of Habitat for Late-Successional and Old-Growth Forest Related Species within the Range of the Northern Spotted Owl.

SOCATS (Southern Oregon Citizens Against Toxic Sprays). 1983. *SOCATS v. Watt:* The evolution of a court case. *NCAP News* 3 (3): 7–9.

Soltani, Atossa, and Penelope Whitney. 1995. *Cut Waste, Not Trees: How to Save Forests, Cut Pollution and Create Jobs.* Rainforest Action Network.

SRI/Shapiro. 1995. *Plant Survey for RTE Designated Species on the Willow Creek Industrial Park Site in Eugene, Oregon.*

Stallworth, Holly. 1995. The cost-benefit paradigm for environmental protection: An economist's perspective on the methodological, theoretical, and ethical problems. *New Solutions* 6 (1): 14–19.

Stone, Deborah. 1989. Causal stories and the formation of policy agendas. *Political Science Quarterly* 104: 281–300.

Summers, Lawrence. 1991. Memorandum to "Distribution" [in World Bank] re: "GEP," December 12.

Swain, Wayland. 1988. Human health consequences of consumption of fish contaminated with organochlorine compounds. *Aquatic Toxicology* 11: 357–377.

Tanabe, Shinsuke. 1988. PCB problems in the future: Foresight from current knowledge. *Environmental Pollution* 50: 5–28.

Tarkowski, S., and E. Yrjanheikki. 1989. WHO-coordinated intercountry studies on levels of PCDDs and PCDFs in human milk. *Chemosphere* 19 (1–6): 995–1000.

Thatcher, Terence. 1990. Understanding interdependence in the natural environment; Some thoughts on cumulative impact assessment under the National Environmental Policy Act. *Environmental Law* 20: 611–647.

Thomas, Kristin Bryan, and Theo Colborn. 1992. Organochlorine endocrine disrupters in human tissue. In *Chemically-Induced Alterations in Sexual and Functional Development: The Wildlife/Human Connection*, ed. T. Colborn and C. Clement. Princeton Scientific Publishing.

Thornton, Joe, Jack Weinberg, and Jay Palter. 1993. *Transition Planning for the Chlorine Phase-Out: Economic Benefits, Costs, and Opportunities*. Greenpeace.

Turpin, Thomas. 1989. Vegetation management and public issues in the Pacific Northwest. Presentation to Fourth Annual Vegetation Management Workshop, Vancouver.

United Nations. 1990. Resolution 44/225: Large-Scale Pelagic Driftnet Fishing and Its Impact on the Living Marine Resources of the World's Oceans and Seas.

UNEP (United Nations Environment Programme) Environmental Effects Panel. 1991. *Environmental Effects of Ozone Depletion: 1991 Update.*

USAID (U.S. Agency for International Development) and U.S. Environmental Protection Agency. 1990. *Ranking Environmental Health Risks in Bangkok, Thailand.*

USDA (U.S. Department of Agriculture). 1993. *Biologic and Economic Assessment of Methyl Bromide Ban.*

USDA. 1996. *Comprehensive Management Plan for the Hells Canyon National Recreation Area. Draft Environmental Impact Statement.*

USDA. 1997a. *Agricultural Chemical Usage: Vegetables 1996 Summary.*

USDA. 1997b. Memorandum from Lon Hatamiya (Agricultural Marketing Service) to the Deputy Secretary re: Proposed Organic Standards, May 1.

USDA. 1997c. Proposed rule: National Organic Program. 7 *CFR* Part 205, December 16.

USDOE (U.S. Department of Energy). 1995. *Closing the Circle on the Splitting of the Atom: The Environmental Legacy of Nuclear Weapons Production in the United States and What the Department of Energy Is Doing about It*. NTIS N95-23562/8INZ.

USDOE. 1996. *Closing the Circle on the Splitting of the Atom: The Environmental Legacy of Nuclear Weapons Production in the United States and What the Department of Energy Is Doing about It*. NTIS DE96009985INZ.

USEPA (U.S. Environmental Protection Agency). 1982. *Health effects guidance for dacthal* (prepared for N.Y. State Department of Health).

USEPA. 1984a. *Guidance for the Interim Registration of Pesticide Products Containing Daminozide as the Active Ingredient.*

USEPA. 1984b. *Memorandum* from John Melone, Hazard Evaluation Division, to Phil Gray, Office of Pesticide Programs regarding criteria for determining which inert ingredients are of toxicological concern and should be given priority review.

USEPA. 1985a. *DCPA (Dacthal) fact sheet.*

USEPA. 1985b. *Report on the Incineration of Liquid Hazardous Wastes by the Environmental Effects, Transport, and Fate Committee, Science Advisory Board.*

USEPA. 1985c. *Work/quality Assurance Project Plan for the Bioaccumulation Study.*

USEPA. 1987a. *Dacthal health advisory.*

USEPA. 1987b. *Unfinished Business: A Comparative Assessment of Environmental Problems. Overview report.*

USEPA. 1989. *Review of OSW's [Office of Solid Waste's] Proposed Controls for Hazardous Waste Incinerators: Products of Incomplete Combustion.*

USEPA. 1990. *Reducing Risk: Setting Priorities and Strategies for Environmental Protection.*

USEPA. 1992. *Framework for Ecological Risk Assessment*. EPA/630/R-92/001.

USEPA. 1993a. Effluent limitations guidelines, pretreatment standards, and new source performance standards: Pulp, paper, and paperboard category; National emission standards for hazardous air pollutants for source category: Pulp and paper production. *Federal Register* 58 (241): 66078–66216.

USEPA. 1993b. 40 *CFR* Parts 63 and 430. Effluent limitations guidelines, pretreatment standards, and new source performance standards: Pulp, paper, and paperboard category; national emission standards for hazardous air pollutants for source category: Pulp and paper production; Proposed rule. *Federal Register* 58 (241): 66078–66214.

USEPA. 1993c. Memorandum from Hugh Kaufman (Hazardous Site Control Division) to Carol Browner (Administrator) re: Recommendation for removal of key EPA officials, immediate shutdown of noncompliant cement kiln hazardous waste burners, creation of national enforcement task force, Inspector General investigation, and reopening of cement kiln hazardous waste burning regulations.

USEPA. 1994a. Risk Characterization of Dioxin and Related Compounds. Draft, May 2.

USEPA. 1994b. Health Assessment Document for 2,3,7,8-tetrachlorodibenzo-p-dioxin (TCDD) and Related Compounds. Review draft.

USEPA. 1994c. Health Assessment Document for 2,3,7,8-Tetrachlorodibenzo-p-dioxin (TCDD) and Related Compounds. Vol. III. Draft.

USEPA. 1994d. *Pesticides Industry Sales and Usage. 1992 and 1993 Market Estimates.*

USEPA. 1994e. *Reregistration Eligibility Decision (RED): Oryzalin.* EPA 738-R-94-016r.

USEPA. 1994f. *Environmental Protection Agency—Office of Pesticides Program Annual Report 1994.* EPA 735-R-95-001.

USEPA. 1998a. Cleaner Technologies Substitutes Assessment for Professional Fabric Care Processes. EPA/7440B-98-001.

USEPA. 1998b. U.S. High Production Volume (HPV) Chemical Hazard Data Availability Study.

USFS. 1993a. *Environmental Assessment for the Hells Canyon Overlook Ii Recreation Development.*

USFS. 1993b. *Viability Assessments and Management Considerations for Species Associated with Late-successional and Old-growth Forests of the Pacific Northwest.*

USFS. 1994a. *Final Supplemental Environmental Impact Statement on Management of Habitat for Late-successional and Old-growth Forest Related Species Within the Range of the Northern Spotted Owl.*

USFS. 1994b. Forest Ecosystem Management: An Ecological, Economic and Social Assessment.

USFWS (U.S. Fish and Wildlife Service). 1992. *Recovery Plan for the Eastern Timber Wolf.*

USFWS. 1994. Biological opinion on the effects of concentration of 2,3,7,8-tetrachlorodibenzo-p-dioxin, to be attained through implementation of a total maximum daily load, on bald eagles along the Columbia River, January 6.

Vail, D. 1994. Symposium introduction: Management of semiarid rangelands—impacts of annual weeds on resource values. In *Ecology and Management of Annual Rangelands: Proceedings*, ed. S. Monsen and S. Kitchen. U.S. Forest Service General Technical Report INT-GTR-313.

Vallette, Jim. 1995. *Deadly Complacency: US CFC Production, the Black Market, and Ozone Depletion.* Ozone Action.

Van Strum, Carol, and Paul Merrell. 1988. Greenpeace comments on the USEPA proposal to amend dioxin and furan policy—53 *Federal Register* 2441. Greenpeace USA.

Van Strum, Carol, and Paul Merrell. 1989. *The Politics of Penta.* Greenpeace USA.

VETC (Virtual Elimination Task Force of International Joint Commission). 1991. *Persistent Toxic Substances: Virtually Eliminating Inputs to the Great Lakes*.

Wapensky, L. 1969. Collaborative study of gas chromatographic and infrared methods for dacthal formulations. *Journal of the Association of Official Analytical Chemists* 52 (6): 1284–1292.

Warshall, Peter. 1994. The biopolitics of the Mt. Graham red squirrel (*Tamiasciuris hudsonicus grahamensis*). *Conservation Biology* 8 (4): 977–988.

Wartenburg, Daniel and Charon Chess. 1993. The risk wars: Assessing risk assessment. *New Solutions* Winter: 16–25.

Webster, R., and H. Maibach. 1985. In vivo percutaneous absorption and decontamination of pesticides in humans. *Journal of Toxicology and Health* 16: 25–37.

Webster, Thomas, and Barry Commoner. 1994. Overview: The dioxin debate. In *Dioxins and Health*, ed. A. Schecter. Plenum.

White, Richard. 1995. *The Organic Machine*. Hill and Wang.

Whiteside, Thomas. 1979. *The Pendulum and the Toxic Cloud: The Course of Dioxin Contamination*. Yale University Press.

Wiles, Richard, and Christopher Campbell. 1993. *Pesticides in Children's Food*. Environmental Working Group.

Wiles, Richard, Kert Davies, and Christopher Campbell. 1998. *Over-Exposed: Organophosphate Insecticides in Children's Food*. Environmental Working Group.

Williams,Terry Tempest. 1994. *An Unspoken Hunger*. Random House.

Wilson, E. 1988. The current state of biological diversity. In *Biodiversity*, ed. E. Wilson. National Academy Press.

Wing, Steve, and Carl Shy. 1992. [Commentary on Wing et al. 1991]. *Environmental Health Monthly* 5 (3): 1.

Wing, S., C. Shy, J. Wood, S. Wolf, D. Cragle, and E. Frome. 1991. Mortality among workers at Oak Ridge National Laboratory: Evidence of radiation effects in follow-up through 1984. *Journal of the American Medical Association* 265 (11): 1397–1402.

Winner, Langdon. 1986. *The Whale and the Reactor: A Search for Limits in an Age of High Technology*. University of Chicago Press.

WMO (World Meteorological Organization). 1994. *Scientific Assessment of Ozone Depletion: 1994, Executive Summary*.

Working Group on Community Right-to-Know. 1993. The Toxics Release Inventory. *Working Notes* (January-February), p. 6.

Wright, Joanne. 1993. Maquiladoras: The cost of women's work. *Free Trade in Perspective* 3 (2): 13.

Index